5G and Beyond Wireless Networks

5G and Beyond Wireless Networks: Technology, Network Deployments, and Materials for Antenna Design offers a comprehensive overview of 5G and beyond 5G wireless networks along with emerging technologies that support the design and development of wireless networks. It also includes discussions on various materials used for practical antenna design which are suitable for 5G, beyond 5G applications, and cell-free massive MIMO systems.

The book discusses the latest techniques used in 5G and beyond 5G (B5G) communication, such as non-orthogonal multiple access (NOMA), device-to-device (D2D) communication, 6G ultra-dense O-RAN, rate-splitting multiple access (RSMA), simultaneous wireless information and power transfer (SWIPT), massive multiple input multiple output (mMIMO), and cell-free massive MIMO systems, which are explained in detail for 5G and beyond cellular networks. The description of NOMA and their benefit for 5G and beyond networks is also addressed along with D2D communication for next generation cellular networks. RSMA technique is also explained for 6G communication. Detailed descriptions for the design and development of 5G and beyond networks over various techniques are included. The materials specification to design antenna for 5G application are also given.

The role of metalens in designing effective antennas and material specifications for 5G applications is explained in this book. Apart from the above emerging topics, this book also gives ideas about intelligent communication, Internet of Multimedia Things (IOMT), millimeter-wave MIMO-UFMC, and fog computing cloud networks. The last chapter gives details about the legal frameworks for 5G technology for responsible and sustainable deployment. Overall, this book may benefit network design engineers and researchers working in the area of next generation cellular networks.

The contents of this book will be helpful for young researchers and master students, and network design engineers who are working in the area of next generation cellular networks.

Materials, Devices, and Circuits: Design and Reliability

Series Editor: Shubham Tayal, K. K. Paliwal, Amit Kumar Jainy

This series is designed to illustrate the many new and exciting challenges in the expansive and interdisciplinary field of materials science, device engineering, reliability, device-circuit co-design and their applications. The scope of this series is broad, with reference works and textbooks offering insight into all aspects of materials development, device fabrication, circuit analysis and their reliability. The titles in this series deliver authoritative information to professionals, researchers, and students across material and device engineering as well as other scientific disciplines. Each volume offers a comprehensive approach covering fundamentals, technologies, their evolution advancements and applications and includes real-world examples where appropriate.

Tunneling Field Effect Transistors
Design, Modeling and Applications
Edited by T.S. Arun Samuel, Young Suh Song, Shubham Tayal, P. Vimala and Shiromani Balmukund Rahi

Quantum-Dot Cellular Automata Circuits for Nanocomputing Applications
Edited by Trailokya Sasamal, Hari Mohan Gaur, Ashutosh Kumar Singh, Xiaoqing Wen

Device Circuit Co-Design Issues in FETs
Edited by Shubham Tayal, Billel Smaani, Shiromani Balmukund Rahi, Samir Labiod, Zeinab Ramezani

Human-Machine Interface Technology Advancements and Applications
Edited by Edited By Ravichander Janapati, Usha Desai, Shrirang Ambaji Kulkarni, Shubham Tayal

Negative Capacitance Field Effect Transistors: Physics, Design, Modeling and Applications
Edited by Young Suh Song, Shubham Tayal, Shiromani Balmukund Rahi, Abhishek Kumar Upadhyay

Laser Applications in Manufacturing
Edited by Pankaj Kumar, Manowar Hussain, Amit Kumar Jain, Sunil Pathak

5G and Beyond Wireless Networks
Technology, Network Deployments and Materials for Antenna Design
Edited by Indrasen Singh, Shubham Tayal, Niraj Pratap Singh, Vijay Shanker Tripathi, and Ghanshyam Singh

For more information about this series, please visit: https://www.routledge.com/Materials-Devices-and-Circuits/book-series/MDCDR

5G and Beyond Wireless Networks

Technology, Network Deployments, and Materials for Antenna Design

Edited by
Indrasen Singh, Shubham Tayal, Niraj Pratap Singh,
Vijay Shanker Tripathi, and Ghanshyam Singh

CRC Press
Taylor & Francis Group
Boca Raton London New York

CRC Press is an imprint of the
Taylor & Francis Group, an **informa** business

Designed cover image: www.shutterstock.com

First edition published 2024
by CRC Press
2385 NW Executive Center Drive, Suite 320, Boca Raton FL 33431

and by CRC Press
4 Park Square, Milton Park, Abingdon, Oxon, OX14 4RN

CRC Press is an imprint of Taylor & Francis Group, LLC

ISBN: 978-1-032-50480-3 (hbk)
ISBN: 978-1-032-52682-9 (pbk)
ISBN: 978-1-003-40783-6 (ebk)

DOI: 10.1201/9781003407836

Typeset in Times LT Std
by KnowledgeWorks Global Ltd.

Contents

Preface

Wireless communication has significantly impacted the world of the digital era because of its enormous applications in many ways. Day to day, the role of wireless communication in our daily lives has increased since its inception. Since 2011 onwards, wireless technology has become an essential part of our lives to make our many tasks quicker and easier. This technology reaches throughout the world, even in some of the more remote regions of the planet. Keeping this in mind, the present book aims to give a basic idea of how to model and implement wireless communication technology such as 5G and beyond 5G cellular networks, their challenges, and future opportunities for research.

Various latest techniques used in 5G and beyond 5G communication, such as non-orthogonal multiple access (NOMA), device-to-device (D2D) communication, 6G ultra-dense O-RAN, rate-splitting multiple access (RSMA), simultaneous wireless information and power transfer (SWIPT), massive multiple input multiple output (mMIMO), and cell-free massive MIMO systems are explained in detail for 5G and beyond cellular networks. Design and development aspects for 5G and beyond networks over various techniques are also included in this book. The different emerging research areas in 5G and beyond wireless networks, probable solutions, and future research directions have been elaborated.

The role of metalens in designing effective antennas and material specifications for 5G applications has been explained in this book. Apart from the above emerging topics, this book also gives ideas about intelligent communication, Internet of Multimedia Things (IOMT), millimeter-wave MIMO-UFMC, and fog computing cloud networks. The last chapter gives details about the legal frameworks for 5G technology for responsible and sustainable deployment. Overall, this book may benefit network design engineers and researchers working in the area of next generation cellular networks.

About the Editors

Dr. Indrasen Singh is presently working as Assistant Professor (Sr. Grade-2) in the Department of Embedded Technology under School of Electronics Engineering, Vellore Institute of Technology, Vellore, Tamil Nadu, India. He received his B. Tech. and M. Tech. degree in Electronics and Communication Engineering from Uttar Pradesh Technical University, Lucknow, India in 2006, and 2010, respectively. He obtained his Ph.D. degree in Electronics and Communication Engineering from National Institute of Technology Kurukshetra, Haryana, India in 2019. He has more than 15 years of teaching/research experience in various reputed technical institutes/ universities. He is supervising B. Tech, M. Tech, and Ph.D. students. He is the reviewer of many SCI/SCOPUS indexed international journals and served as TPC member and reviewer in different conferences. He has published more than 20 research papers in national/international journals/conferences of repute and many are under review. His research interests are in the area of cooperative communication, stochastic geometry, modeling of wireless networks, heterogeneous networks, millimeter wave communications, device-to-device communication, and 5G/6G communication.

Dr. Shubham Tayal is Assistant Professor in the Department of Electronics and Communication Engineering at SR University, Warangal, India. He has more than six years of academic/research experience of teaching at UG and PG level. He has received his Ph.D. in Microelectronics & VLSI Design from National Institute of Technology, Kurukshetra, M.Tech (VLSI Design) from YMCA University of Science and Technology, Faridabad and B.Tech (Electronics and Communication Engineering) from MDU, Rohtak. He has qualified GATE (2011, 2013, 2014) and UGC-NET (2017). He has published more than 45 research papers in various international journals and conferences of repute and many papers are under review. He is on the editorial and reviewer panel of many SCI/SCOPUS indexed international journals and conferences. Currently, he is editor/co-editor of eight books. He acted as keynote speaker and delivered professional talks on various forums. He is a member of various professional bodies like IEEE, IRED, etc. He is on the advisory panel of many international conferences. He is a recipient of Green ThinkerZ International Distinguished Young Researcher Award 2020. His research interests include simulation and modeling of multi-gate semiconductor devices, device-circuit co-design in digital/analog domain, machine learning, and IoT.

Dr. Niraj Pratap Singh is an Associate Professor in the Department of Electronics and Communication Engineering at National Institute of Technology Kurukshetra, India. He received his B.E. and M.E. degrees in Electronics and Communication Engineering from Birla Institute of Technology, Mesra, Ranchi, India in 1991 and 1994, respectively. He received the Ph.D. degree in Electronics and Communication Engineering from National Institute of Technology, Kurukshetra, India. His current research interests are in the area of wireless communications including stochastic geometry, device-to-device communication, radio resource management, cognitive radio, and 5G communications.

Prof. Vijay Shanker Tripathi has received his B.Tech. degree from J.K. Institute of Applied Physics and Technology, University of Allahabad in 1988. He did his M.Tech and Ph.D. from Motilal Nehru National Institute of Technology Allahabad in 1999 and 2007, respectively. At present he is a professor in the Department of Electronics and Communication Engineering at MNNIT Allahabad. He has published more than 70 research papers in reputed refereed international journals as well as in international conferences. He has guided seven Ph.D. scholars and more than 15 M.Tech scholars. His area of specialization includes antenna, SDR, and non-invasive RF sensors.

Prof. Ghanshyam Singh is a Professor of Electronics and Communication Engineering and Director of the Centre for Smart Information and Communication Systems, Department of Electrical and Electronic Engineering Science at the University of Johannesburg, South Africa. He received PhD degree in Electronics Engineering from the Indian Institute of Technology, Banaras Hindu University, Varanasi, India, in 2000. He was a Visiting Researcher at the Seoul National University, Seoul, South Korea. He has worked as a Professor with the Department of Electronics and Communication Engineering, Jaypee University of Information Technology, Wakanaghat, Solan, India. He is an author/co-author of more than 370 scientific research papers of peer reviewed Journal and International Conferences. His research and teaching interests include broad range of spectrum (RF/Microwave, Millimeter/THz Wave, Free-Space Optics and Visible Light) Technologies with it's emerging Applications such as in Next-Generation Communication Systems (5G/6G), Sustainable Smart City, Industry 4.0/5.0, Healthcare 4.0, Intelligent Transport System, Energy Management (IoE) and Digital Farming. He has more than 23 years of teaching and research experience from Academia and Research & Development Institutions. He has supervised various doctoral thesis and Master's dissertations. He has executed several sponsored research projects of ISRO, DRDO and currently working on 5G Enabling Technology Localization Resource Allocation sponsored by SENTECH. He is the author of several books and book chapters published by Springer, Elsevier, IET, Wiley and CRC.

Contributors

Mehboob Ul Amin
Department of Electrical and Electronic
   Engineering Science
University of Johannesburg, APK Campus
Johannesburg, South Africa

Rabindra K Barik
KIIT University
Bhubaneswar, Odisha, India

Bikash Ranjan Behera
Department of Electronics and
   Communication Engineering
Vel Tech Rangarajan Dr. Sagunthala R&D
   Institute of Science and Technology
Chennai, Tamil Nadu, India

Abhijit Bhowmick
Department of CE, SENSE
Vellore Institute of Technology (VIT)
Vellore, Tamil Nadu, India

Kaibalya Prasad Bhuyan
National Institute of Technology
Rourkela, Odisha, India

Ram Chandra Singh Chauhan
Department of Electronics and
   Communication Engineering
Institute of Engineering and Technology
Lucknow, Uttar Pradesh, India

Gopal Chandra Das
Department of ECE, NIT
Patna, India

Abhilasha Gautam
Symbiosis Institute of Technology,
   Symbiosis International (Deemed
   University)
Pune, Maharashtra, India

Aman Jolly
KIET Group of Institutions
Ghaziabad, Uttar Pradesh, India

Rebant Juyal
Assam University
Silchar, Assam, India

Mohd Javed Khan
Dept. of ECE, Integral University
Lucknow, Uttar Pradesh India

Abhishek Kumar
CentraleSupelec
Universite Paris-Saclay
Rennes, France

Dileep Kumar
Department of Electronics and
   Communication Engineering
National Institute of Technology
Kurukshetra, India

Sumit Kundu
Department of Electronics and
   Communication Engineering
National Institute of Technology
Durgapur, India

Vinay Kumar
Department of Electronics and
   Communication Engineering
Motilal Nehru National Institute of
   Technology Allahabad
Prayagraj, Uttar Pradesh, India

V. Krishnaveni
Dept. of ECE, PSG College of
   Technology
Coimbatore, Tamil Nadu, India

Sumit Kushwaha
University Institute of Computing
Chandigarh University
Chandigarh, India

Soumya Ranjan Mahapatro
School of Electronics Engineering
(SENSE)
Vellore Institute of Technology (VIT)
Chennai, Tamil Nadu, India

Vikas Pandey
Babu Banarasi Das University
Lucknow, Uttar Pradesh, India

Kirti Prakash
Department of Electronics and
Communication Engineering
Institute of Engineering and Technology
Lucknow, Uttar Pradesh, India

Manish Sabraj
School of Electrical Engineering
Shri Mata Vaishno Devi University
(SMVDU)
Katra, Jammu and Kasmir, India

Seemanti Saha
Department of ECE, NIT
Patna, India

Hritwika Sarkar
Cadence Design Systems Ltd
Noida, Uttar Pradesh, India

Ghanshyam Singh
Department of Electrical and Electronic
Engineering Science
University of Johannesburg, APK Campus
Johannesburg, South Africa

Gyanesh Singh
School of Electrical Engineering
Shri Mata Vaishno Devi University
(SMVDU)
Katra, Jammu and Kasmir, India

Indrasen Singh
School of Electronics Engineering
Vellore Institute of Technology (VIT)
Vellore, Tamil Nadu, India

Shashibhushan Sharma
Department of Electronics and
Communication
University of Allahabad
Prayagraj, Uttar Pradesh, India

Niraj Pratap Singh
Department of Electronics and
Communication Engineering
National Institute of Technology
Kurukshetra, India

Dheeraj Kumar Sharma
Electronics and Communication
Engineering
National Institute of Technology
Kurukshetra
Kurukshetra, Haryana, India

Prabhat Thakur
Symbiosis Institute of Technology
Symbiosis International (Deemed
University)
Pune, Maharashtra, India

E. Udayakumar
Dept. of ECE
Kalaignarkarunanidhi Institute of
Technology
Coimbatore, Tamil Nadu, India

Gaurav Verma
Department of Electronics and
Communication Engineering
National Institute of Technology
Kurukshetra, India

1 Machine Learning Empowered Resource Allocation to Reconfigurable Intelligent Surface Supported *MIMO-NOMA Networks*

Mehboob Ul Amin and Ghanshyam Singh

1.1 INTRODUCTION

The radio spectrum is a valuable and limited resource that needs to be utilized effectively. Cognitive radio (CR) and non-orthogonal multiple access (NOMA) are emerging technologies that hold promise for improving spectrum utilization and providing seamless communication with higher efficiency, lower power consumption, improved accuracy, and reduced computational complexity. With the exponential growth of wireless technologies, the deployment of 5G networks and research on 6G technologies are underway to achieve more reliable and faster communication. In this context, intelligent reflecting surface (IRS) has emerged as a key technique for next-generation networks, including beyond 5G (B5G) and 6G communication networks. IRS comprises a large number of passive elements that can reflect signals with adjustable phase shifts. These elements are typically made of metallic materials. The beam steering capability of IRS allows it to direct the signal toward the intended user, enhancing the signal quality and coverage. The main objective of modelling an IRS system is to maximize the utilization of the reflected wave, which directly influences the IRS's performance. By optimizing the phase shifts of the reflecting elements, the IRS can effectively enhance signal strength, improve channel capacity, and optimize overall system performance. Moreover, IRS technology, with its beam steering capability and adjustable phase shifts, has the potential to significantly improve the performance of future communication networks such as B5G and 6G. By maximizing the utilization of the reflected waves, IRS can enhance signal quality and coverage, leading to improved spectral efficiency, lower power consumption, and more accurate communication with reduced computational complexity.

DOI: 10.1201/9781003407836-1

A CR is an enhanced version of a software radio that incorporates a cognitive engine consisting of a knowledge base, reasoning engine, and learning engine. This cognitive engine enables software alterations based on inference derived from the information stored in the knowledge base, which serves as the radio's permanent memory. The reasoning engine functions as an intelligent system, as commonly referred to in artificial intelligence literature. The learning engine is responsible for updating the knowledge base through experiences and observations. In wireless communication systems, the allocation of resources such as power, bandwidth, and spectrum play a crucial role in determining the system's performance. Effective resource allocation is vital for maximizing the operation of next-generation communication systems. Therefore, resource allocation has become a prominent feature of future communication networks. To enhance the rate and efficiency of wireless channels, software-controlled meta-surfaces, known as IRS, or reconfigurable intelligent surfaces (RIS), or large intelligent surface/antenna (LISA), can be deployed. These surfaces can be implemented using large arrays of cost-effective antennas with inter-distances approximately half the wavelength or through the use of meta-material elements with significantly smaller sizes and inter-distances compared to the wavelength. The input to the IRS consists of plane waves, and the output consists of scattered waves whose phase shifts are controlled to achieve the desired reflection. However, IRS has certain limitations, such as its inability to support high digital processing capabilities, as it is primarily designed based on the concept of beamforming. Therefore, other complementary concepts may need to be employed alongside IRS to overcome these limitations.

1.1.1 RELATED WORK

The integration of IRS into the CR system aims to improve spectral efficiency, energy efficiency, and overall system performance. By optimizing the beamforming vector at the secondary user (SU) transmitter and phase shifts at the IRS, the study demonstrates significant improvements in achievable rates and system capacity. In [1], the authors propose an innovative approach to enhance the performance of a multiple-input single-output (MISO)-CR system by incorporating IRS. The results highlight the potential of IRS as a valuable technology for enhancing the performance of CR systems in dynamic spectrum access scenarios. This system optimizes the beamforming vector at the SU transmitter and phase shifts at the IRS to improve the SU's achievable efficiency. The study considers power capacity limits on the SU transmitter, interference temperature bounds on the primary user (PU) receivers, spectrum efficiency (SE), and energy efficiency (EE). Numerical results demonstrate that the proposed IRS significantly enhances the SU's achievable rate in both ideal and imperfect channel state information (CSI) scenarios. Similarly [2] also focuses on enhancing SE and EE in a CR system using the same approach as in [1]. Computational experiments are conducted to evaluate the potential gains in SU's achievable rate. In the context of multiple-input multiple-output (MIMO)-CR systems, [3] suggests a similar principle can be applied. To solve the complex optimization problem with variable coupling, the block coordinate descent (BCD) technique is introduced. Resource allocation strategies within CR systems are discussed in reference [4]. The optimization problem is formulated as non-convex, considering the lack of PU knowledge of the CSI. The BCD technique is employed to address this problem. In the presence of significant cross-link interference with the PU, reference [5] proposes the use of an IRS to facilitate spectrum sharing between the PU and SU. However, security concerns such as main user emulation, jamming, eavesdropping, and spectrum sensing data falsification can pose challenges in CR networks. [6] focuses on increasing the secrecy rate of the SU under various constraints. This approach ensures the protection of legitimate users' data from eavesdropping, ensuring communication anonymity. In CR networks, beamforming and pre-coding techniques offer advantages in terms of SE, power control, link capacity, transmission security, and improved signal-to-interference plus noise ratio (SINR). [7] investigates the use of IRS and vertical beamforming in CR networks to maximize spectrum utilization. [8] introduces an IRS to enhance the network throughput of CR systems. The study addresses beamforming design considering bounded CSI error and statistical CSI error models for PU-related channels in IRS-aided CR systems. Finally [9] presents multiuser underlay cognitive transceiver topologies that aim to provide mm-wave spectrum access while reducing interference to existing users. Overall, the mentioned references explore various aspects of CR systems [10–12], including the use of IRS [13–15], optimization techniques for resource allocation [16–18], beamforming strategies [19, 20], and addressing security and interference challenges in CR networks [21, 22]. In this chapter, we describe a proposed approach using machine learning and deep learning techniques for an IRS-based NOMA cognitive system, focusing on enhancements in ergodic successive convex approximation, ensuring optimal solutions for the transmit vector of base station and IRS vector. The deep learning solution is implemented for higher gains of capacity and target rate at optimal signal-to-noise ratio (SNR). The main contribution of this chapter is as follows:

- We propose a new machine learning–based deep learning approach for an IRS-based NOMA system to get enhancements in ergodic capacity and target rate.
- For better optimization of resources, we use first-order Taylor decomposition to solve the non-convexity problem of sum rates defined by Shannon theorem and convert the non-convex problem as the difference of two convex functions by successive convex approximation.
- Deep learning solution–based artificial neural network using MATLAB software is proposed for better performance gains. Simulation results depict that proposed deep learning solution increases the spectral efficiency in terms of bps/Hz and lowers the outage probability (OP) and computational complexity.

The rest of the chapter is organized as follows: Section 1.2 presents the system model, providing an explanation of the key components and their interactions. In Section 1.3, the problem formulation is described, outlining the specific objectives and constraints for achieving optimal solutions, and Section 1.4 gives us the optimal solution. Section 1.5 presents the results and discussions, analyzing the performance and effectiveness of the proposed approach. Finally, in Section 1.6, a summary is provided, highlighting the main findings and conclusions of the study.

1.2   SYSTEM MODEL

We consider an RIS-assisted multiuser downlink network, with base station equipped with multiple antennas and RIS with M passive reflecting elements to enhance the channel. The RIS's diagonal reflection network for the n-th block is given

$$\theta[n] = diag(e^{j\theta_1[n]}, e^{j\theta_2[n]}, \cdots e^{j\theta_M[n]}). \tag{1.1}$$

The phase shift of the mth passive reflecting element is designated by θ_m, wherein the significance of θ_m is between 0 and 2π and n ~ CN $(0,\sigma^2)$ designates the AWGN (additive white Gaussian noise).The link between base station (BS) and IRS is modeled as Rayleigh, and Rican channel is used between IRS and user equipment's. Let r_m denote the channel gain between BS and IRS reflecting element and r'_m denote the channel between m-th IRS reflecting element and n-th user. There may be certain elements that carry data in terms of diagonal matrix. Let d_r denote the diagonal matrix carrying data between BS and m-th IRS reflecting element, the received SINR is given as

$$SINR = \frac{\left|d_r^H \theta[n]d_r\right|^2 P_x}{\displaystyle\sum_{i=n+1}^{N} \left|d_{r,x}^H \theta[n]\right|^2 P_i + N^2}, \tag{1.2}$$

where P_x is the power at transmitter. The NOMA user rate is given as

$$C = log_2(1 + SINR). \tag{1.3}$$

For N number of users, the total sum rate is given as

$$CSum = \sum_{n=1}^{N} log_2(1 + SINR). \tag{1.4}$$

1.3   PROBLEM FORMULATION

The following logarithm concave problem is formulated at IRS and can be solved using sub-gradient iterative algorithm involving deep learning–based ML algorithm

$$\max \theta[n] \sum_{n=1}^{N} log_2(1 + SINR),$$

$$\text{s.t } |\theta_m|^2 = 1, \forall m \in \{1,2,...,M\}, \tag{1.5}$$

$$\theta_1 + \theta_2 + \cdots + \theta_n = 1$$

We introduce the variable T_n for target rate such that

$$\left(1 + \min\left(SINR\right)\right) \geq T_n. \tag{1.6}$$

Thus, we have,

$$\sum C = log2\left(\prod_{n \in N} T_n\right) \tag{1.7}$$

While solving the concavity of equation (1.7), we have following sub-problem

$$\max T_n, \tag{1.8}$$

$$s.t \ T^2 \leq \prod_{n \in N} C_n, \tag{1.8-a}$$

$$C - 1 \leq \min(SINR). \tag{1.8-b}$$

For the link between IRS element and users, we introduce IRS reflecting vector R^H and channel coefficient r_m such that the sub-problem of equation (1.8) can be written as

$$\frac{\sum\limits_{n=1}^{N} \left|R^H r_m d_r^H\right|^2}{\sum\limits_{i=1}^{Ni} \left|R^H r_m d_i\right|^2 + N^2} \geq T_n - 1, \tag{1.9}$$

$$= \sum\limits_{Ni}^{i=1} \left|R^H r_m d_i\right|^2 \leq \frac{\sum\limits_{N}^{n=1} \left|R^H r_m d_r^H\right|^2}{T_n - 1}. \tag{1.10}$$

1.4 OPTIMAL SOLUTION

The SINR of equation (1.2) is now formulated as:

$$SINR^* = \frac{P_t \theta[n] \left|R^H r_m d_r^H\right|^2}{P_t \sum\limits_{i=n+1}^{N} \theta[n] \left|R^H r_m d_r^H\right|^2 + N^2}. \tag{1.11}$$

Then, the target rate T_n^* is given as

$$T_n^* = \sum\limits_{n=1}^{N} BW \log_2\left(I + SINR^*\right), \tag{1.12}$$

where I is the identity matrix of receive antennas. Let $\mathcal{R} = RR^H$ and $\boldsymbol{H} = r_m^i d_r^H$, where $\mathcal{R} \geq 0$ and $rank(\mathcal{R}) = 1$, $\boldsymbol{H} > 0$ and $rank(\boldsymbol{H}) = 1$, then T_n^* can be presented as function of $\mathcal{R}$ as

$$T_n^* = \sum_{n=1}^{N} BW\left[f_1(\mathcal{R}) - f_2(\mathcal{R})\right]. \tag{1.13}$$

Now, T_n^* is the difference between two concave functions. The rank one constraint is thus non-concave, which is first transformed as the difference of two convex function as

$$rank(\mathcal{R}) = 1, \tag{1.14}$$

$$tr(\mathcal{R}) - \|\mathcal{R}\|_2 = 0, \tag{1.15}$$

where $tr(\mathcal{R}) = \sum_{n=1}^{N} S_n$ and S_n represents the n-th largest singular value of $\mathcal{R}$. $\|\mathcal{R}\|_2$ is the spectral norm of matrix $\mathcal{R}$. We use successive convex approximation to replace norm $\|\mathcal{R}\|_2$ with its first-order Taylor approximation for obtaining lower bound values as

$$\|\mathcal{R}\|_2 \geq \|\mathcal{R}^i\|_2 tr\left(d_r^i d_r^i\left(\mathcal{R} - \mathcal{R}^i\right)\right) \cong \|\bar{\mathcal{R}}\|_2. \tag{1.16}$$

After adding the penalty term $\eta \gg 0$, the target rate can be formulated as

$$T_n^* = \max_{R} \sum_{n=1}^{N} BW\left[f_1(\mathcal{R}) - \overline{f_2(\mathcal{R})}\right] - \eta\left(tr(\mathcal{R}) - \|\bar{\mathcal{R}}\|_2\right). \tag{1.17}$$

1.4.1 DEEP LEARNING SOLUTION

This solution is based on deep learning neural networks, where we consider an input layer, a hidden layer, and an outer layer. The input layer size is determined by the number of RIS reflecting elements (M) and number of NOMA users. The activation function "tanh" is selected to match the input and output relationships. The RIS phase-shift matrix $\theta[n]$ and channel matrix gain $\boldsymbol{H}$ determine the size of input layer so that the input to DNN network is $\theta[n]$ and $\boldsymbol{H}$. The DNN will determine current RIS matrix $\theta[n]$ and corresponding beamforming vector $\mathcal{R}$ as follows:

$$\{\theta[n], \mathcal{R}\} = \psi\left(\theta[n-1], \boldsymbol{H}\right), \tag{1.18}$$

where ψ represents the DNN-fitted function. This action achieves the target rate T_n and can be fed with the loss function $L(n)$. Thus, optimization problem can be solved as

$$L(n) = -\sum_{n=1}^{N}\left(\overline{T_n} + \overline{H}\theta[n]\right)^2. \tag{1.19}$$

The addition of penalty term gives the following expression for loss function as

$$L(n) = -\overline{T_n} + W\left\|\psi\right\|_2^2. \tag{1.20}$$

1.5 RESULTS AND DISCUSSIONS

This section evaluates the performance of a deep learning (DL)–based RIS-NOMA system in terms of ergodic capacity (EC), target rate, OP, symbol error rate (SER), and computational complexity using Monte Carlo simulations carried on MATLAB software. Three users are selected in the NOMA system and are designated as D1, D2, and D3. D3 is the strong user, D2 is the intermediate user, and D1 belongs to the weaker user. The parameters are set in terms of bandwidth BW = 10 MHZ, carrier frequency=3 GHZ, noise variance=−94 dbm, distance from BS to D1 = 1000 meters, distance from BS to D2=750 meters, distance from BS to D3=500 meters, and NOMA coefficients $\alpha1=0.74$, $\alpha2=0.5$, and $\alpha3=0.25$.

Figure 1.1 shows the plot for EC vs. source transmits power for NOMA, IRS-NOMA, and DL-based IRS-NOMA for three users D1, D2, and D3. The EC starts increasing with the increase in source transmit power for all the users. IRS-NOMA for all the three users has much greater performance than NOMA. The application of DL further enhances the capacity for all three users with DL IRS NOMA

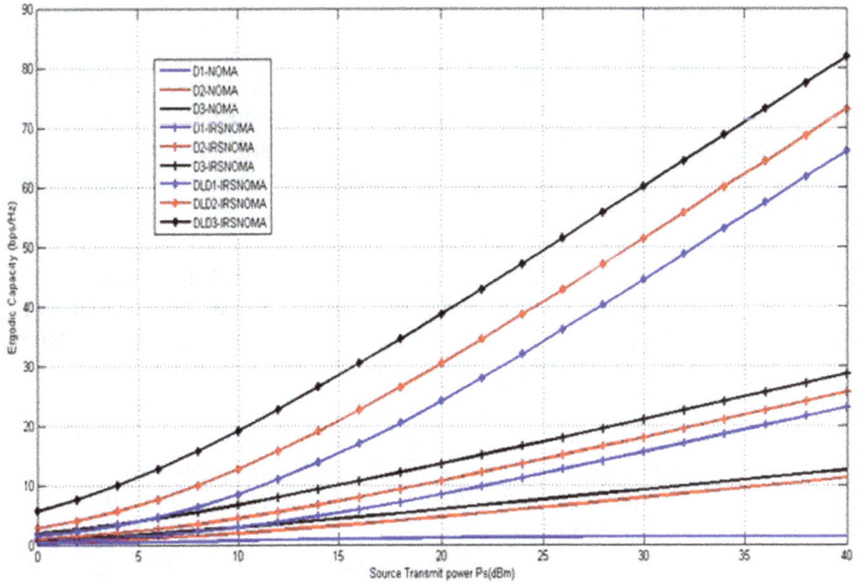

FIGURE 1.1 Ergodic capacity vs. source transmits power.

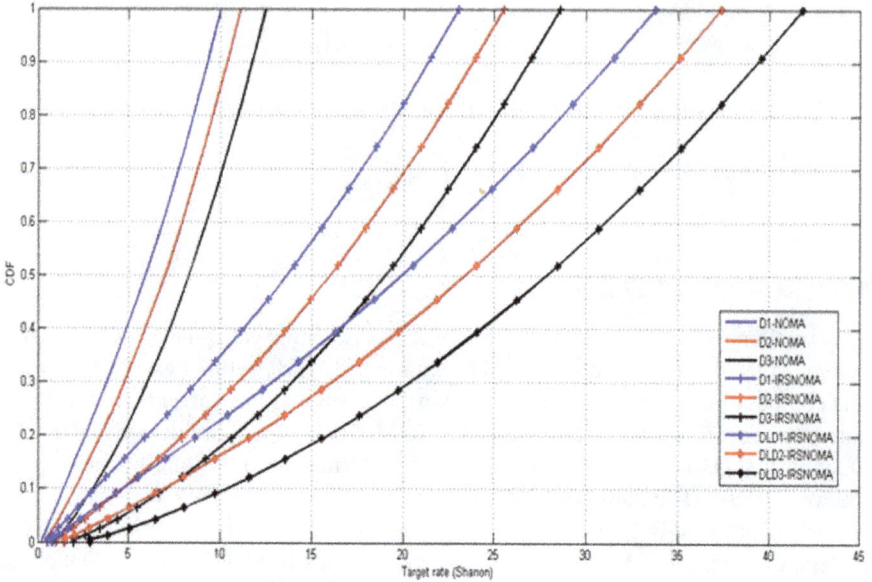

FIGURE 1.2 CDF plot of target rate.

for D3 having ergodic capacity of 81 bps/Hz at the source transmit power of 40 dBm. Better gains are being achieved by DL-based IRS NOMA as compared to IRS NOMA. The IRS NOMA achieves the better results as compared to NOMA. The IRS NOMA for D3 position is having EC of 28 bps/Hz, while NOMA for D3 position is having ergodic capacity of 11 bps/Hz. Similarly, the weak user D1 also exhibits better capacity gains with the application of the DL algorithm. DL-based IRS-NOMA has ergodic capacity of 67 bps/Hz for D1 user, while IRS NOMA has the capacity of 22 bps/Hz and NOMA has a capacity of 1 bps/Hz at 40 dBm source transmit power. Figure 1.2 shows the cumulative distribution function (CDF) plot of target rate in terms of unit of capacity known as Shannon for all the proposed schemes. CDF plot is done in order to reflect the performance of cell edges and cell centers. Two rate measures 5th and 50th percentile are used to reflect the coverage performances of weak and strong users, respectively. NOMA for all the three users has poor performance at 5th percentile level CDF (i.e., at cell edges). The target rate for NOMA starts increasing at higher percentile levels, NOMA having a target rate of 7 Shannon at 50th percentile level for D3 user. The incorporation of IRS also does not have sufficient impact on the target rate for 5th percentile level CDF. However, the target rate increases for the 50th percentile level CDF with IRS NOMA having target rate of 19 Shannon for D3 user. The DL-based IRS NOMA has some impact on the target rate for the 5th percentile level CDF, with DL-based IRS NOMA having a target rate of 7 Shannon for D3 user. Similarly, for the 50th percentile CDF level, DL-based IRS NOMA has the highest target rate of about 27 Shannon.

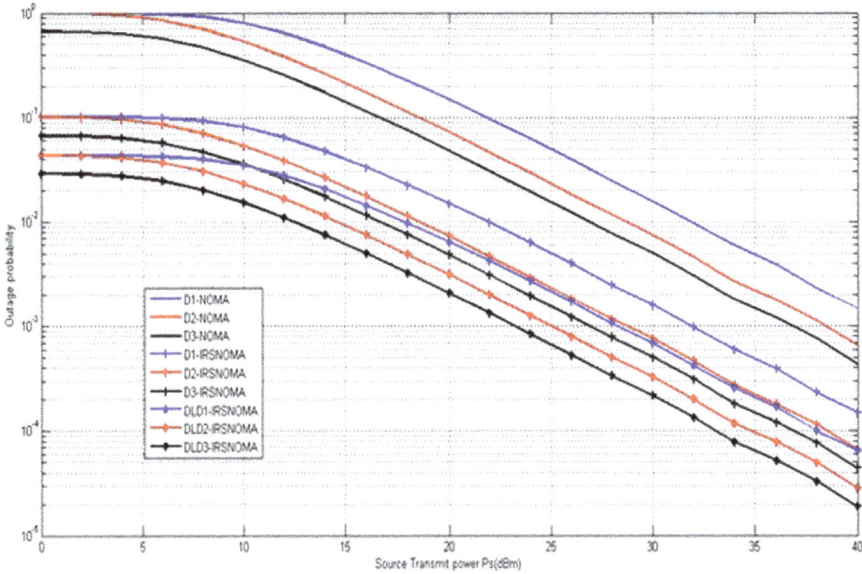

FIGURE 1.3 Plot for outage probability vs. source transmit power.

Figure 1.3 illustrates the OP of NOMA, IRS NOMA, and DL-based IRS NOMA for all three users. The OPs with DL-based IRS NOMA and IRS NOMA are gradually lower than those of NOMA. The OP of 10^{-2} is achieved by NOMA for D3 user at the source transmit power of 30 dBm. IRS NOMA for the same user achieves the same outage at 20 dBm, while DL-based IRS NOMA achieves the same OP for the same user at 12 dBm. Thus, there is 8 dBm improvement in power while using DL-based IRS NOMA. In this way, power can be optimized using DL-based IRS NOMA.

Figure 1.4 shows the symbol error rate vs. SNR for all the proposed techniques. The graph illustrates that DL-based IRS NOMA lowers the SER for all the users. NOMA has the highest SER, and SER shows a gradual decrease with the incorporation of IRS and DL-based IRS NOMA. Finally, we calculate the computational complexity of NOMA, IRS NOMA, and DL-based IRS NOMA in terms of time taken by each algorithm in Figure 1.5 while evaluating SNR. DL-based IRS NOMA for D3 user has the lowest computational complexity of almost 0.028 (ms), while DL-based IRS NOMA has the highest computational complexity of 0.075 (ms).

1.6   CONCLUSION

In this chapter, we have introduced a novel DL-based solution for an IRS system to enhance the ergodic capacity and target rate of NOMA users. Through extensive simulations using MATLAB software, we have validated the analytical analysis

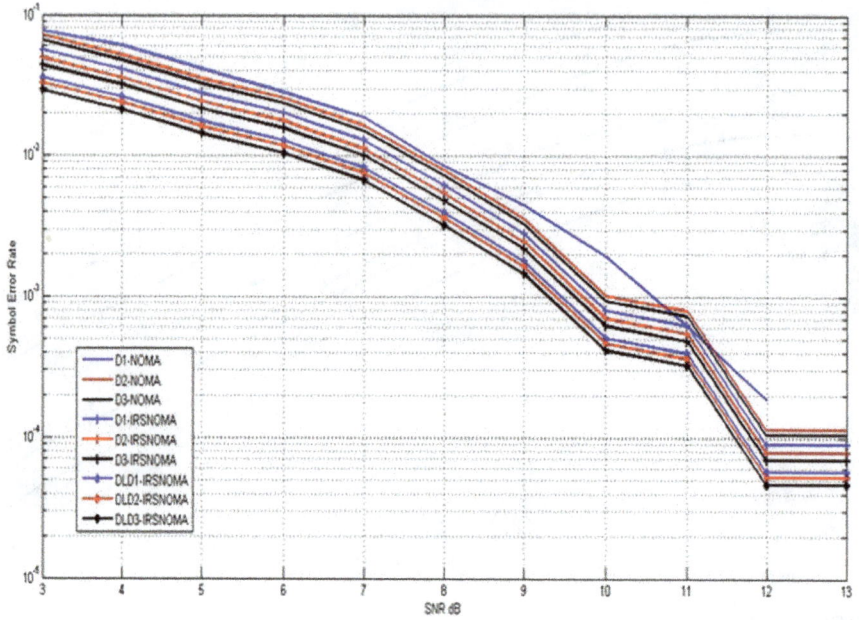

FIGURE 1.4   SER vs. SNR of proposed NOMA systems.

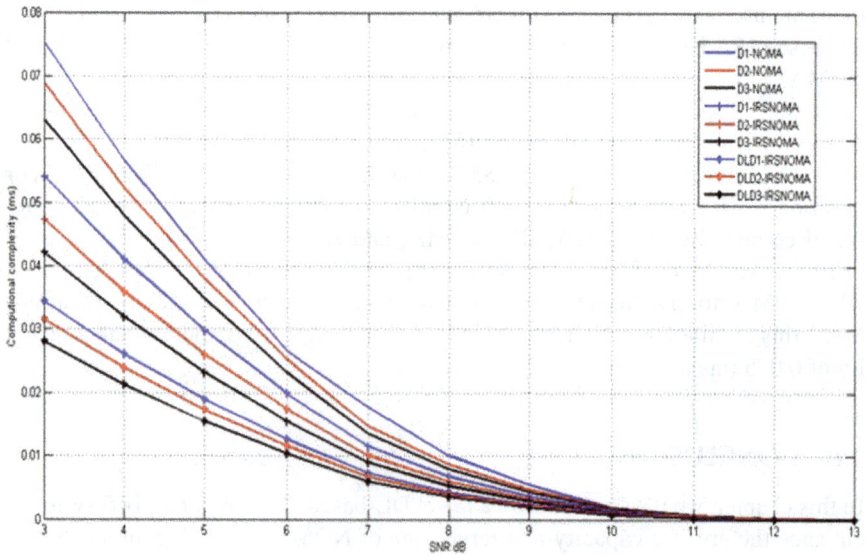

FIGURE 1.5   Plot for computational complexity vs. SNR.

and demonstrated the effectiveness of our proposed method. By employing the proposed DL approach, we have achieved significant improvements in reducing the SER of NOMA systems while maintaining low computational complexities. The simulation results have shown enhanced gains in terms of ergodic capacity, outage probability, and target rate, aligning with the requirements of 5G and B5G networks. Furthermore, with the modifications introduced in our system model based on DL techniques, it can also be applied to localization in the context of IoT and 5G technologies. This opens up new possibilities for leveraging the advantages of deep learning in optimizing and enhancing various aspects of wireless communication systems.

REFERENCES

1. X. Guan, Q. Wu, and R. Zhang, "Joint power control and passive beamforming in IRS-assisted spectrum sharing," *IEEE Communications Letters*, vol. 24, no. 7, pp. 1553–1557, 2020.
2. L. Dong, H.-M. Wang, and H. Xiao, "Secure cognitive radio communication via intelligent reflecting surface," *IEEE Transactions on Communications*, vol. 69, no. 7, pp. 4678–4690, 2021.
3. S. F. Zamanian, S. M. Razavizadeh, and Q. Wu, "Vertical beamforming in intelligent reflecting surface-aided cognitive radio networks," *IEEE Wireless Communications Letters*, vol. 10, no. 9, pp. 1919–1923, 2021.
4. L. Zhang, C. Pan, Y. Wang, H. Ren, and K. Wang, "Robust beamforming design for intelligent reflecting surface aided cognitive radio systems with imperfect cascaded CSI," *IEEE Transactions on Cognitive Communications and Networking*, vol. 8, no. 1, pp. 186–201, 2022.
5. V. Kakkad, M. LeFevre, K. Roy Choudhury, J. Kisslo, and G. E. Trahey, "Effect of transmit beamforming on clutter levels in transthoracic echocardiography," *Ultrasonic Imaging*, vol. 40, no. 4, pp. 215–231, 2018.
6. A. Alkhateeb, S. Alex, P. Varkey, Y. Li, Q. Qu, and D. Tujkovic, "Deep learning coordinated beamforming for highly-mobile millimeter wave systems," *IEEE Access*, vol. 6, pp. 37328–37348, 2018.
7. S. Kutty and D. Sen, "Beamforming for millimeter wave communications: An inclusive survey," *IEEE Communications Surveys & Tutorials*, vol. 18, no. 2, pp. 949–973, 2016.
8. C. G. Tsinos, S. Chatzinotas, and B. Ottersten, "Hybrid analog-digital transceiver designs for multi-user MIMO mmwave cognitive radio systems," *IEEE Transactions on Cognitive Communications and Networking*, vol. 6, no. 1, pp. 310–324, 2020.
9. C. Kumar Sheemar and D. Slock, "Hybrid beamforming for bidirectional massive MIMO full duplex under practical considerations," *Proceeding of the 93rd IEEE Vehicular Technology Conference (VTC2021-Spring)*, 25–28 April 2021, Helsinki, Finland, pp. 1–5.
10. H. Mu and T. Hu, "Cognitive Radio and the New Spectrum Paradigm for 5G," Matin, M. (eds) "Spectrum Access and Management for Cognitive Radio Networks," Signals and Communication Technology. Springer: Singapore, 2017. https://doi.org/10.1007/978-981-10-2254-8_10
11. D. Wang, F. Zhou, W. Lin, Z. Ding, and N. Al-Dhahir, "Cooperative hybrid nonorthogonal multiple access-based mobile-edge computing in cognitive radio networks," *IEEE Transactions on Cognitive Communications and Networking*, vol. 8, no. 2, pp. 1104–1117, 2022.
12. A. Srivastava, M. S. Gupta, and G. Kaur, "Energy efficient transmission trends towards future green cognitive radio networks (5G): Progress, taxonomy and open challenges," *Journal of Network and Computer Applications*, vol. 168, pp. 102760/1-32, 2020.

13. J. Singh, I. Chatterjee, S. Srivastava, and A. K. Jagannatham, "Hybrid transceiver design and optimal power allocation in downlink mmwave hybrid MIMO cognitive radio systems," *Proceeding of the National Conference on Communications (NCC)*, 24–27 May 2022, Mumbai, India, pp. 1–6.

14. R. Alhamad, "Nonorthogonal multiple access with adaptive transmit power and energy harvesting using intelligent reflecting surfaces for cognitive radio networks," *Signal, Image and Video Processing*, vol. 17, pp. 83-89, 2023.

15. Z. Lin, M. Lin, B. Champagne, W.-P. Zhu, and N. Al-Dhahir, "Robust hybrid beamforming for satellite-terrestrial integrated networks," *Proceeding of the IEEE International Conference on Acoustics, Speech and Signal Processing (ICASSP 2020)*, 04–08 May 2020, Barcelona, Spain, pp. 8792–8796.

16. J. Du, W. Xu, Y. Deng, A. Nallanathan, and L. Vandendorpe, "Energy-saving UAV-assisted multiuser communications with massive MIMO hybrid beamforming," *IEEE Communications Letters*, vol. 24, no. 5, pp. 1100–1104, 2020.

17. J. Mansukhani and P. Ray, "Censored spectrum sharing strategy for MIMO systems in cognitive radio networks," *IEEE Transactions on Wireless Communications*, vol. 18, no. 12, pp. 5500–5510, 2019.

18. I. Ahmed, H. Khammari, A. Shahid, A. Musa, K. S. Kim, E. De Poorter, and I. Moerman, "A survey on hybrid beamforming techniques in 5G: Architecture and system model perspectives," *IEEE Communications Surveys & Tutorials*, vol. 20, no. 4, pp. 3060–3097, 2018.

19. G. Hegde, Y. Cheng, and M. Pesavento, "Hybrid beamforming for large-scale MIMO systems using uplink-downlink duality," *Proceeding of the IEEE International Conference on Acoustics, Speech and Signal Processing (ICASSP 2017)*, 05–09 March 2017, New Orleans, LA, USA, pp. 3484–3488.

20. H. S. Antony and T. Lakshmanan, "Secure beamforming in 5G-based cognitive radio network," *Symmetry*, vol. 11, no. 10, pp. 1260/1-15, 2019.

21. A. F. Tayel, A. H. El-Malek, S. I. Rabia, and A. M. Abdelrazek, "Securing hybrid channel access cognitive radio networks with energy harvesting," *Physical Communication*, vol. 45, pp. 101260/1-15, 2021.

22. H. Bany Salameh, R. Tashtoush, H. Al-Obiedollah, A. Alajlouni, and Y. Jararweh, "Power allocation technique with soft performance guarantees in hybrid OFDMA–NOMA cognitive radio systems: Modeling and simulation," *Simulation Modelling Practice and Theory*, vol. 112, pp. 102370/1-12, 2021.

2 Cooperative Power-Domain Non-Orthogonal Multiple Access (NOMA) in 5G Systems
Potentials and Challenges

Mohd Javed Khan and Indrasen Singh

2.1 INTRODUCTION

Many cutting-edge technologies have recently been presented and investigated in order to improve the efficiency of fifth-generation (5G) networks. NOMA is one of the primary and key multiple access techniques (MAT) in the 5G wireless network [1]. MATs have played a critical role in the evolution of mobile communication systems. The first-generation (1G) mobile communication systems, such as advanced mobile phone system (AMPS) and total access communication system (TACS), used frequency division multiple access (FDMA) as the primary multiple access technique [2]. In FDMA, each user is assigned a unique frequency band for transmitting their data. The second-generation (2G) mobile communication systems used global system for mobile communications (GSM) and code division multiple access (CDMA) as its MAT [3], in which 2G used time division multiple access (TDMA) and CDMA as the primary MATs for GSM and CDMA [4, 5], respectively. In TDMA, the available bandwidth is divided into time slots, and each user is assigned a unique time slot for transmitting their data. In CDMA, each user is assigned a unique code sequence for transmitting their data. The third-generation (3G) mobile communication systems, such as universal mobile telecommunications system (UMTS) and code division multiple access 2000 (CDMA2000), used CDMA as the primary multiple access technique. CDMA2000 also used time division-synchronous code division multiple access (TD-SCDMA) as an additional multiple access technique. The fourth-generation (4G) mobile communication systems, such as long-term evolution (LTE) [6] and worldwide interoperability for microwave access (WiMAX), used orthogonal frequency-division multiple access (OFDMA) [7] as the primary multiple access technique. In OFDMA, the available bandwidth is divided into subcarriers, and each user is assigned a subset of subcarriers for transmitting their data. The fifth-generation (5G) mobile communication system uses a combination of OFDMA and NOMA as the primary MAT [8]. In NOMA, multiple users can share the

DOI: 10.1201/9781003407836-2

same subcarrier by using different power levels for their transmissions. The choice of MAT in mobile communication systems depends on several factors, including the number of users, the available bandwidth, the desired data rate, and the power constraints. The evolution of mobile communication systems has seen a shift toward more efficient and flexible MATs, enabling higher data rates and improved system performance.

2.1.1   NOMA

NOMA [9] is a wireless MAT that enables multiple users to share the same frequency and time resources. The fundamental idea behind NOMA is to allow multiple users to transmit their signals simultaneously on the same channel, which improves spectral efficiency and increases capacity.

In traditional orthogonal multiple access (OMA) techniques, different users are assigned separate time or frequency resources, and their signals are transmitted separately. This restricts the number of users who can be served concurrently. NOMA, on the other hand, allows several users to share the same resources by allocating various power levels to each user's signal. In other words, stronger users are assigned lower power levels, while weaker users are assigned higher power levels.

The key to successful NOMA is to exploit the different channel conditions of the users. By assigning different power levels, the stronger user's signal can be detected at a lower power level, while the weaker user's signal can be detected at a higher power level. This allows multiple users to use the same resources simultaneously, resulting in higher spectral efficiency and capacity. NOMA and OMA are two different MATs used in wireless communication systems. Here are some differences between NOMA and OMA on the basis of the literature survey [10]:

a. Spectrum efficiency: NOMA is a more spectrum-efficient technology than OMA, because it enables numerous users to utilize the same band of frequency, but with different codebooks, power levels, or other parameters, which increases the capacity of the network. In contrast, OMA allocates different frequency bands to different users, which can lead to inefficient use of spectrum resources.

b. Number of users: NOMA can support a larger number of users compared to OMA. This is because NOMA allows multiple users to share the same resources, while OMA allocates different resources to different users that may restrict the number of users who can be supported.

c. Fairness: OMA provides a fair distribution of resources among users, while NOMA can prioritize certain users over others. NOMA can allocate more resources to users with higher quality of service requirements, which can lead to a less fair distribution of resources.

d. Complexity: NOMA is generally more complex than OMA, as it requires more sophisticated signal processing and resource allocation techniques to manage the multiple users sharing the same resources. In contrast, OMA is a simpler technology that is easier to implement.

e. Interference: NOMA can have higher levels of interference between users sharing the same resources, which can affect the overall network performance. In contrast, OMA can have lower interference, as each user is allocated different resources.

Therefore, NOMA and OMA are MATs that offer different benefits and drawbacks. NOMA is a more spectrum-efficient technology that can support a larger number of users, but can also be more complex and less fair. OMA is a simpler technology that provides a fair distribution of resources among users, but can be less spectrum-efficient and support fewer users.

There are two different implementations of NOMA: time-domain NOMA (TD-NOMA) and power-domain NOMA (PD-NOMA). In TD-NOMA, different users transmit their signals in different time slots, while in PD-NOMA, different power levels are assigned to different users. PD-NOMA is the more commonly used variant because it is easier to implement and offers better performance in practical scenarios.

NOMA is a promising technology for various applications, including machine-to-machine (M2M) communication, Internet of Things (IoT), and 5G networks. NOMA can provide significant improvements in spectral efficiency and capacity, which are crucial for meeting the growing demands of wireless communications.

2.1.1.1 Time-Domain NOMA

In TD-NOMA [11], the base station (BS) assigns different time slots to different users based on their channel conditions. Those who have better channel conditions are given shorter time slots, whereas those who have weaker channel conditions are given longer time periods in NOMA system. The users transmit their signals in their respective time slots, and the BS employs advanced signal processing techniques to separate and recover the signals at the receiver.

TD-NOMA relies on the fact that different users experience different channel conditions, such as different path loss, fading, and interference. By assigning different time slots to different users, TD-NOMA can exploit the diversity in the channel conditions to achieve higher spectral efficiency. However, TD-NOMA requires accurate channel estimation and synchronization [12], and it may suffer from interference between users in neighboring time slots. To overcome these challenges, advanced signal processing techniques, such as multiuser detection, interference cancellation, and beamforming, can be employed. For a variety of applications, TD-NOMA has been suggested as a viable technique including M2M communication, IoT, and 5G networks. Here are some benefits of TD-NOMA [11]:

a. Increased capacity: TD-NOMA can increase the capacity of the wireless system. This can result in a higher number of supported users and better utilization of the spectrum resources [12].
b. Fairness: TD-NOMA can provide fairness among users as each user is allocated different power levels at different time instances, which can help ensure a more equitable distribution of resources.

c. Low latency: TD-NOMA can provide low latency as users can transmit data simultaneously in the same time-frequency resources. Applications that demand real-time communication, such as online gaming and video conferencing, may benefit from this.

d. Reduced interference: TD-NOMA can reduce interference among users, as the allocation of different power levels can help mitigate the effects of inter-user interference. This can improve the overall network performance and user experience.

e. Energy efficiency: TD-NOMA can be more energy-efficient than other MATs, as it allows for a higher number of users to share the same resources and it decreases the overall energy consumption of the network [13].

TD-NOMA is a promising technology that can provide several benefits for wireless communication networks, including increased capacity, fairness, low latency, reduced interference, and energy efficiency. These benefits make TD-NOMA a useful technology for supporting emerging applications that require high data rates, low latency, and reliable connectivity.

2.1.1.2 Power-Domain NOMA (PD-NOMA)

PD-NOMA [14] is a NOMA variant that enables several users to share the same frequency band by employing various power levels to distinguish their transmissions. In PD-NOMA, the BS assigns various users varying power levels depending on the channel state. Lower power levels are assigned to users who have better channel state, and higher power levels are assigned to users who have worse channel state.

PD-NOMA relies on the fact that different users experience different channel conditions, such as different path loss, fading, and interference. By allocating different power to different users, PD-NOMA can exploit the diversity in the channel conditions to achieve higher spectral efficiency and capacity. The BS uses advanced signal processing techniques such as successive interference cancellation (SIC) to separate and recover the signals from different users.

PD-NOMA can improve the spectral efficiency, energy efficiency, and overall performance of wireless communication systems, especially in dense and highly dynamic environments. PD-NOMA can also improve the fairness of the system by allowing weaker users to use the same resources as stronger users, providing equal opportunities to all users. PD-NOMA can be used in various wireless communication scenarios, including cellular networks, M2M communication, and IoT. Here are some benefits of PD-NOMA [15]:

a. Increased spectral efficiency: PD-NOMA can considerably improve the spectral efficiency of a cellular system by enabling several users to share the same resources. This means that more users can be supported in a given bandwidth, leading to a higher data rate and better overall system performance.

b. Fairness: PD-NOMA assigns various users varying power levels depending on the channel state. This allows for a more equitable distribution of resources among users, which can improve overall system fairness.

c. Low latency: PD-NOMA can also reduce latency in a wireless communication system, as it allows for more efficient use of resources. By transmitting multiple messages in the same resource block, PD-NOMA can reduce the amount of time needed to transmit each message, leading to lower latency and better overall system performance [16].

d. Improved reliability: PD- NOMA can also improve the reliability of a wireless communication system. By using different power levels for different users, it can provide better resistance to fading and other channel impairments, resulting in a more reliable and robust system.

e. Energy efficiency: PD- NOMA can also improve energy efficiency in wireless communication systems. PD-NOMA can reduce the total power consumption of the system, leading to longer battery life and lower energy costs.

PD-NOMA is a promising technique that can offer significant benefits for wireless communication systems, including increased spectral efficiency, fairness, low latency, improved reliability, and energy efficiency.

PD-NOMA has some challenges, including channel estimation, interference between users, and increased complexity due to the use of advanced signal processing techniques. However, these challenges can be addressed by designing efficient algorithms and protocols and by employing advanced signal processing techniques.

Overall, PD-NOMA is a promising technology that can address the challenges associated with the increasing demand for wireless communication and the limited availability of spectrum resources

2.1.1.3 Comparison of PD NOMA and TD NOMA

PD-NOMA and TD-NOMA [17] are two variants of NOMA that allow multiple users to share the same frequency range by allocating different resources to differentiate between their signals. While both PD-NOMA and TD-NOMA have similar objectives, there are some key differences between these two techniques.

i. **Resource allocation:** In PD-NOMA, different power levels are allocated to different users to differentiate between their signals, while in TD-NOMA, different time slots are allocated to different users. PD-NOMA uses PD multiplexing, while TD-NOMA uses TD multiplexing.

ii. **Channel astimation:** In PD-NOMA, accurate channel estimation is required to assign different power levels to different users, while in TD-NOMA, accurate channel estimation is required to assign different time slots to different users. Channel estimation can be challenging in both techniques, but the channel estimation requirements are different.

iii. **Interference:** PD-NOMA can experience interference between users, especially when the users have similar power levels or overlapping coverage areas. TD-NOMA can also experience interference, especially when users have similar channel conditions or are located in close proximity.

iv. **Complexity:** PD-NOMA requires sophisticated signal processing techniques such as SIC and power allocation algorithms, which increase the complexity of the system. TD-NOMA also requires complex signal processing techniques, such as superposition coding and decoding, which can increase the complexity of the system.

v. **Fairness:** PD-NOMA allows weaker users to use the same resources as stronger users, improving the fairness of the system. TD-NOMA can also improve fairness by allowing weaker users to use the same time slots as stronger users.

vi. **Deployment:** PD-NOMA may require new hardware and infrastructure to be deployed, while TD-NOMA can be implemented using existing hardware and infrastructure.

Overall, both PD-NOMA and TD-NOMA have their advantages and disadvantages, and the choice between these techniques depends on the specific requirements and constraints of the wireless communication scenario. PD-NOMA is generally more suitable for scenarios where power control is feasible, such as cellular networks, while TD-NOMA is more suitable for scenarios with low power consumption and latency requirements, such as M2M communication and IoT applications.

2.1.2   INTEGRATION OF NOMA

Recently, NOMA has been studied and integrated with various other technologies to enhance their performance [18]. Here are a few examples of NOMA integration with other technologies:

i. **NOMA and massive MIMO:** Massive MIMO involves using a large number of antennas at the transmitter and receiver to increase the number of spatial dimensions in the communication channel. This allows for multiple users to be served simultaneously, which increases the capacity and efficiency of the wireless network. Massive MIMO also improves the reliability of wireless communication by reducing interference and increasing the signal-to-noise ratio. NOMA can be integrated with massive MIMO technology to further increase the spectral efficiency and capacity of cellular networks. In this integration, multiple users are served simultaneously using the same time-frequency resources by exploiting the large number of antennas in massive MIMO BSs [19].

ii. **NOMA and millimeter-wave (mm-wave) technology:** mm-wave technology involves using frequencies in the mm-wave band, which is in the range of 30–300 GHz, for wireless communication. This frequency range provides high bandwidth, which allows for high data rates and low latency. However, mm-wave signals are highly directional and have limited range, which requires the use of beamforming and other advanced techniques to ensure reliable communication. NOMA can be combined with mm-wave technology to increase the data rate and capacity of wireless networks. This integration can provide high throughput and low latency communication. NOMA can be used to increase the number of users that can be supported in a given frequency band, while mm-wave technology can be used to provide high data rates and low latency [20].

iii. **NOMA and cooperative communication:** NOMA can be integrated with cooperative communication to improve the reliability and coverage of wireless networks. In this integration, the users in the network can help each other to transmit their data by relaying each other's signals, leading to increased coverage and reliability [21].

iv. **NOMA and cloud radio access network (C-RAN):** C-RAN is a centralized architecture for wireless networks where the baseband processing functions are performed at a centralized cloud computing platform, while the remote radio heads (RRHs) handle the radio frequency (RF) functions. This architecture enables the efficient allocation of resources, centralized management and optimization of the network, and the ability to support a large number of users and devices. NOMA can be integrated with C-RAN technology to enhance the efficiency of resource allocation and reduce the computational complexity of the BS. In this integration, the BS can be divided into a remote radio head and a central processing unit, with the NOMA transmission being processed in the central unit. C-RAN is a centralized architecture for wireless networks where the baseband processing functions are performed at a centralized cloud computing platform, while the RRHs handle the RF functions. This architecture enables the efficient allocation of resources, centralized management and optimization of the network, and the ability to support a large number of users and devices [22].

There are several other potential integration scenarios for NOMA with other technologies, and ongoing research is exploring the possibilities and evaluating their benefits. Cooperative power-domain non-orthogonal multiple access (CPD-NOMA) is one of them. Most investigations in the literature have focused on CD-NOMA [19–23] or PD-NOMA [20, 24], though without user or device cooperation. To the best of our knowledge, there has been no systematic study or review of CPD-NOMA. In this chapter we study it in detail.

2.2 COOPERATIVE POWER-DOMAIN NOMA (CPD-NOMA)

When several wireless devices collaborate to send and receive data, this sort of wireless communication is referred to as cooperative communication. In this type of communication, wireless devices cooperate with each other to improve the overall quality of the communication link and increase the reliability of data transmission. Cooperative wireless communication can be implemented in various ways, such as cooperative relaying, cooperative diversity, and cooperative beamforming. Cooperative relaying involves using multiple wireless devices to transmit data between a source and destination, with each device transmitting a part of the data. Cooperative diversity involves using multiple antennas on each wireless device to improve the quality of the communication link. Cooperative beamforming involves directing the transmission of data toward the receiver using multiple antennas and signal processing techniques. The benefits of cooperative wireless communication include increased range, improved reliability, and better quality of service. It can also help reduce interference and improve the efficiency of wireless networks. Cooperative wireless communication is particularly useful in situations where the wireless signal is weak or there are obstacles in the transmission path. It can be used in various applications, such as cellular networks, wireless sensor networks, and ad-hoc networks. Therefore, cooperative wireless communication is an important technique for improving the performance and reliability of wireless communication, and it has many potential applications in various domains.

CPD-NOMA is a wireless communication technology that allows multiple users to share the same frequency and time resources. CPD-NOMA combines two key concepts: PD-NOMA and cooperative communications. In CPD-NOMA, the BS assigns different power levels to each user on the basis of their channel conditions, allowing weaker users to use the same resources as stronger users. Figure 2.1 shows the arrangement of CPD-NOMA [23]. CPD-NOMA results in higher spectral efficiency, better energy efficiency, and lower latency compared to traditional OMA systems.

The cooperative aspect of CPD-NOMA involves the use of multiple antennas or relays to improve the system performance. Cooperative beamforming [24] and relay-assisted CPD-NOMA are two examples of cooperative strategies that can be used to enhance the performance of CPD-NOMA systems. These cooperative strategies can further increase the capacity, reliability, and coverage of CPD-NOMA systems, making them a promising technology for future wireless communication networks.

CPD-NOMA systems can provide higher spectral efficiency, better energy efficiency, and lower latency compared to traditional communication technologies. With the increasing demand for high-speed data transmission and the emergence of new applications such as M2M communication, IoT, and 5G networks, CPD-NOMA is a leading technique for future cellular systems.

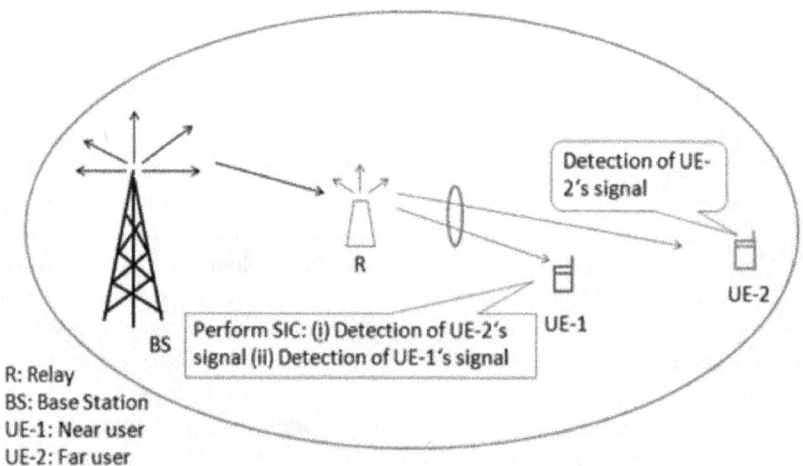

FIGURE 2.1   Illustration of downlink CPD-NOMA transmission.

CPD-NOMA has some challenges, including the design of efficient cooperation strategies, the coordination among users, and the complexity of the signal processing techniques. However, these challenges can be addressed by developing advanced algorithms and protocols that can facilitate cooperation among users.

2.2.1 ADVANTAGES OF CPD-NOMA

i. **Increased spectral efficiency:** CPD-NOMA can improve the spectral efficiency of the wireless communication system by allowing multiple users to share the same frequency band and by exploiting the diversity in their channel conditions and power resources [25].

ii. **Improved energy efficiency:** CPD-NOMA can reduce the energy consumption of the system by enabling users to share their power resources and reducing the need for users to transmit at high power levels.

iii. **Improved reliability:** CPD-NOMA can enhance the reliability of the system by allowing users to cooperate with each other and improve their transmission and reception capabilities, especially in scenarios where some users have poor channel conditions or limited power resources [26].

iv. **Fairness:** CPD-NOMA can provide equal opportunities to all users by allowing weaker users to cooperate with stronger users to access the same resources.

v. **Flexibility:** CPD-NOMA can be used in different wireless communication scenarios, including cellular networks, M2M, and IoT.

2.2.2 LIMITATIONS OF CPD-NOMA

i. **Complexity:** CPD-NOMA requires advanced signal processing techniques and coordination among users, which increase the complexity of the system.

ii. **Interference:** CPD-NOMA can experience interference between cooperating users, which can affect the performance of the system [27].

iii. **Design challenges:** CPD-NOMA requires efficient cooperation strategies and protocols to facilitate cooperation among users, which can be challenging to design.

iv. **Resource allocation:** CPD-NOMA requires efficient resource allocation algorithms to allocate power levels and channel access among users.

Overall, CPD-NOMA is a promising technology that can improve the performance of the wireless communication system, especially in scenarios where users have heterogeneous channel conditions and power resources. However, CPD-NOMA also has some limitations that need to be addressed to fully exploit its benefits.

2.3 DESIGN PRINCIPLES OF CPD-NOMA SYSTEMS

The power allocation algorithm used in CPD-NOMA–based systems is a crucial component that determines the power levels allocated to each user for transmission. The power allocation algorithm aims to maximize the overall system performance by considering the channel conditions and power resources of each user.

The power allocation algorithm in CPD-NOMA consists of two main steps: user grouping and power allocation.

   i. **User grouping:** The first step of the power allocation algorithm is to group the users into clusters based on their channel states. Users are grouped together when the channel states are optimal, and individually when the channel states are less optimal. This grouping allows the system to exploit the diversity in channel conditions and power resources to improve the overall performance [25].

   ii. **Power allocation:** The second step of the power allocation algorithm is to allocate the power levels to each user for transmission. The power levels are allocated based on the users' channel states and power resources. The users with better channel states and lower power resources are allocated lower power levels, while the users with poorer channel states and higher power resources are allocated higher power levels. This power allocation strategy allows the weaker users to cooperate with the stronger users and access the same resources, improving the overall system performance [27].

The power allocation algorithm in CPD-NOMA can be implemented using different optimization techniques, such as convex optimization, dynamic programming, and heuristic algorithms. The algorithm's complexity depends on the number of users and the channel conditions, and it should be designed to balance the performance and complexity trade-off.

Overall, the power allocation algorithm is a critical component of CPD-NOMA, which plays a crucial role in improving the system's spectral efficiency, energy efficiency, and reliability. The algorithm should be designed carefully to consider the users' channel conditions and power resources and balance the trade-off between complexity and performance.

2.3.1 FACTORS THAT INFLUENCE POWER ALLOCATION IN CPD-NOMA

In CPD-NOMA, the power allocation algorithm plays a crucial role in determining the power levels allocated to each user for transmission. The power allocation algorithm aims to maximize the overall system performance by considering various factors that influence the power allocation decision. The following are some of the factors that influence power allocation in CPD-NOMA:

   i. **Channel conditions:** The channel conditions of each user are a critical factor that influences power allocation in CPD-NOMA. Users with better channel conditions can transmit at lower power levels, while users with poorer channel conditions require higher power levels to achieve the desired performance.

   ii. **Power resources:** The power resources available to each user also influence the power allocation decision. Users with higher power resources can transmit at lower power levels, while users with limited power resources require higher power levels to achieve the desired performance.

iii. **User grouping:** The grouping of users based on their channel conditions is an essential factor that influences power allocation in CPD-NOMA. Users with similar channel conditions are grouped together to exploit the diversity in channel conditions and power resources and improve the overall system performance.

iv. **Quality of service (QoS) requirements:** The QoS requirements of each user also influence power allocation in CPD-NOMA. Users with higher QoS requirements require higher power levels to achieve the desired performance, while users with lower QoS requirements can transmit at lower power levels.

v. **Interference:** The interference between cooperating users is another factor that influences power allocation in CPD-NOMA. Interference can affect the system's performance and should be considered when allocating power levels to each user.

vi. **Energy efficiency:** The energy efficiency of the system is also an important factor that influences power allocation in CPD-NOMA. The power levels allocated to each user should be optimized to reduce the energy consumption of the system while maintaining the desired performance.

The power allocation algorithm in CPD-NOMA should consider all these factors to allocate the power levels efficiently and optimize the overall system performance. The algorithm's design should balance the trade-off between performance and complexity and should be adapted to the specific wireless communication scenario.

2.3.2 IMPACT OF POWER ALLOCATION ON SYSTEM PERFORMANCE

Power allocation is a critical factor that impacts the performance of CPD-NOMA systems. The power allocation algorithm aims to allocate the power levels efficiently to each user for transmission, optimizing the overall system performance. The following are some ways in which power allocation affects the system performance:

i. **Spectral efficiency:** The spectral efficiency of the system is directly related to the power allocation decision. The power levels allocated to each user affect the transmission rate. By allocating power levels efficiently, the spectral efficiency of the system can be increased, allowing more users to access the same resource and increasing the overall throughput of the system.

ii. **Energy efficiency:** The power allocation algorithm also affects the energy efficiency of the system. By allocating power levels efficiently, the energy consumption of the system can be reduced, increasing the system's energy efficiency. Efficient power allocation also leads to a longer battery life for wireless devices, reducing the need for frequent recharging.

iii. **System reliability:** The power allocation algorithm affects the reliability of the system by ensuring that each user can transmit at the desired power level without interference from other users. By allocating power levels efficiently, the system can achieve high reliability, reducing the probability of packet loss, and increasing the overall quality of service.

iv. **Fairness:** Power allocation also affects the fairness of the system by ensuring that each user can access the same resources fairly. By allocating power levels efficiently, the system can achieve high fairness, allowing each user to access the same resources, and reducing the probability of congestion or delay.

v. **Interference management:** Power allocation also plays a critical role in managing interference between users. By allocating power levels efficiently, the interference between cooperating users can be minimized, allowing the users to transmit at the desired power level without affecting each other's performance [25].

In summary, the power allocation algorithm plays a critical role in determining the system's performance in CPD-NOMA. By allocating power levels efficiently, the system's spectral efficiency, energy efficiency, reliability, fairness, and interference management can be optimized, improving the overall system performance.

2.4   COOPERATIVE STRATEGIES FOR CPD-NOMA SYSTEMS

Cooperative strategies in wireless communication refer to the techniques that enable wireless devices to work together to improve the overall system performance. In CPD-NOMA systems, cooperative strategies involve collaboration between multiple users to improve the spectral and energy efficiency of the system. The following are some of the cooperative strategies used in CPD-NOMA [28]:

i. **Cooperative relaying:** In this strategy, one user acts as a relay to help another user transmit its data to the BS. By relaying the data, the transmission becomes more reliable and can overcome the effects of fading and interference.

ii. **Joint decoding:** In this strategy, users with strong channels can help users with weak channels by jointly decoding their signals at the receiver. This allows the weak user to transmit at a lower power level, saving energy and increasing the overall spectral efficiency.

iii. **User clustering:** In this strategy, users with similar channel conditions are grouped together to form clusters. Each cluster is assigned a power level based on the channel conditions of the users in the cluster, maximizing the overall spectral efficiency of the system.

iv. **Interference alignment:** In this strategy, users coordinate their transmissions to align their interference with each other, reducing the overall interference in the system. This results in a higher spectral efficiency and improved energy efficiency.

Cooperative strategies in CPD-NOMA can significantly improve the system performance by allowing users to work together to optimize the spectral and energy efficiency of the system. These strategies can be implemented using distributed algorithms, allowing the users to collaborate without the need for a central controller. By leveraging the benefits of cooperation, CPD-NOMA can achieve higher spectral and energy efficiency compared to traditional wireless communication techniques.

2.5   SOME IMPORTANT TECHNIQUE FOR CPD-NOMA SYSTEMS

2.5.1   COOPERATIVE BEAMFORMING TECHNIQUE FOR CPD-NOMA SYSTEMS

Beamforming [29] for CPD-NOMA is a technique used in cellular communication systems to enhance spectral efficiency and increase the capacity of the system. In CPD-NOMA, two or more users share the same frequency band and time slot, but they are allocated different power levels according to their respective channel conditions.

Beamforming is used to focus the transmitted signal toward the intended receiver(s) by adjusting the phasing and amplitude of the signal at different antennas in the transmitter. By doing so, the signal power can be directed toward the intended receiver(s) and away from other users, reducing interference and improving the system's overall performance.

CPD-NOMA also involves cooperation among the users, where stronger users help weaker users achieve reliable communication by relaying their signals. By using beamforming in conjunction with cooperative NOMA, the system can achieve higher throughput, lower latency, and improved reliability, leading to a better user experience.

2.5.1.1   Benefits of Beamforming for CPD-NOMA Systems

Beamforming in CPD-NOMA systems provides several benefits, such as

    i. **Enhanced channel capacity:** Beamforming helps to direct the signal energy to the intended user and reduces the interference, thereby increasing the capacity of the channel.
    ii. **Improved spectral efficiency:** Beamforming enhances the spectral efficiency of CPD-NOMA systems by enabling multiple users to simultaneously send and receive information within the same frequency band.
    iii. **Better coverage:** The beamforming technique allows for better coverage and reach, which means that users located far away from the BS can still benefit from CPD-NOMA systems.
    iv. **Power efficiency:** Beamforming in CPD-NOMA systems optimizes the power consumption and reduces energy wastage by directing the signal energy toward the intended user.
    v. **Interference reduction:** The beamforming technique minimizes the interference from nearby cells and improves the overall signal quality, thereby providing a better user experience.

In summary, beamforming helps to achieve a high-quality, reliable, and efficient performance of CPD-NOMA systems.

2.5.2 Relay-assisted CPD-NOMA

Relay-assisted CPD-NOMA [30] is a promising technique for improving the performance of wireless communication systems. In this technique, a relay station is used to assist in the transmission of signals between the source and destination nodes, and NOMA is used to allow multiple users to share the time slot and same frequency band simultaneously.

Several studies have investigated the benefits of relay-assisted CPD-NOMA. These studies suggest that relay-assisted cooperative power domain NOMA can improve the performance of wireless communication systems in various ways, making it a promising technique for future wireless networks. On the basis of a literature study of CPD-NOMA, the benefits are classified as

i. **Increased spectral efficiency:** Relay-assisted CPD-NOMA increases the spectral efficiency of the system. This means that more data can be transmitted in a given amount of time, improving the overall capacity of the network. This benefit was also demonstrated in a study, where they showed that relay-assisted CPD-NOMA can achieve higher spectral efficiency compared to conventional cooperative NOMA and conventional PD-NOMA schemes.

ii. **Improved coverage:** The use of relay stations in CPD-NOMA networks can improve the coverage of the system, especially in areas with poor signal strength. The relay station can help extend the range of the signal and reduce the likelihood of dropped calls or lost connections. This benefit was also shown in a study where they demonstrated that relay-assisted CPD-NOMA can provide better coverage compared to conventional cooperative NOMA and conventional PD-NOMA schemes.

iii. **Better reliability:** The use of relay stations in CPD-NOMA networks can also improve the reliability of the system. By having multiple transmission paths, the system is less susceptible to interference, and the probability of transmission errors is reduced. This benefit was demonstrated in a study where they showed that relay-assisted CPD-NOMA can achieve higher reliability compared to conventional cooperative NOMA and conventional power domain NOMA schemes.

iv. **Lower latency:** Relay-assisted cooperative power domain NOMA can also reduce latency in the network. By reducing the number of hops between the source and destination, the transmission delay is minimized, improving the overall user experience. This benefit was also demonstrated in a study, where they showed that relay-assisted cooperative power domain NOMA can achieve lower latency compared to conventional cooperative NOMA and conventional PD-NOMA schemes.

v. **Reduced energy consumption:** Relay-assisted CPD-NOMA can also decrease the energy consumption of cellular devices. By reducing the number of retransmissions and increasing the efficiency of transmissions, it can help in the reduction of energy consumption of cellular devices, making them more sustainable and environmentally friendly. This benefit was demonstrated in a study by Alsharoa et al., where they showed that relay-assisted CPD-NOMA can provide higher energy efficiency compared to conventional cooperative NOMA and conventional PD-NOMA schemes.

2.5.3 JOINT POWER AND RELAY SELECTION ALGORITHMS FOR RELAY-ASSISTED CPD-NOMA

Joint power and relay selection algorithms are important for optimizing the performance of relay-assisted CPD-NOMA. These algorithms aim to select the optimal relay node and power-allocation strategy to maximize the system capacity, minimize the power consumption, or balance the trade-off between these two objectives.

There are several joint power and relay selection algorithms for relay-assisted CPD-NOMA, such as

i. **Greedy algorithm:** This algorithm selects the relay node and power allocation that maximizes the sum rate. It starts by selecting the user with the weakest channel quality, and then iteratively adds the user with the next weakest channel quality until the maximum number of users is reached. The greedy algorithm for CPD-NOMA can be summarized as follows [31]:
   a. Sort the users in descending order of channel strength.
   b. Assign the user with the strongest channel as the primary user.
   c. For each subsequent user, calculate the potential rate gain from using that user as a relay for the primary user.
   d. Select the user with the highest potential rate gain and add it to the set of relays.
   e. Repeat steps c and d until the desired number of relays is reached or no further rate gain can be achieved.
   f. Allocate power to each user in the set of relays such that the total power used for transmission is minimized while meeting the desired QoS requirements for each user.
   g. The greedy algorithm for CPD-NOMA is a suboptimal solution that is easy to implement and has low computational complexity. However, it may not always provide the optimal solution, and there may be cases where the greedy algorithm fails to achieve the desired QoS requirements. In such cases, more advanced optimization techniques, such as dynamic programming or convex optimization, may be necessary to achieve the desired performance.

ii. **Genetic algorithm (GA):** This algorithm uses a genetic search approach to find the optimal relay node and power allocation. It generates a set of candidate solutions and evaluates their fitness based on the system capacity. The best solutions are then selected for the next generation until the optimal solution is found. The basic steps of a GA for CPD-NOMA can be summarized as follows [32]:
   a. Initialize the population: Generate an initial population of candidate solutions, where each solution corresponds to a possible combination of power allocation, user clustering, and relay selection.

b. Evaluate fitness: Evaluate the fitness of each candidate solution based on a fitness function that reflects the system performance in terms of sum rate and QoS requirements.

c. Selection: Select the fittest solutions from the population to generate the parent population for the next generation.

d. Crossover: Apply crossover operations to the parent population to generate offspring solutions.

e. Mutation: Apply mutation operations to the offspring solutions to introduce diversity.

f. Evaluation: Evaluate the fitness of the offspring solutions.

g. Replacement: Select the fittest solutions from the parent and offspring populations to form the new population for the next generation.

h. Termination: Repeat steps c to g until a termination condition is met, such as reaching a maximum number of generations or achieving a desired level of performance.

The GA can be used to optimize the power allocation, user clustering, and relay selection jointly to maximize the system performance in terms of sum rate while satisfying the QoS requirements of each user. The GA can handle complex optimization problems and provide global optimality, but it may have high computational complexity and require a large number of iterations to converge to a solution.

iii. **Particle swarm optimization (PSO):** This algorithm uses a swarm intelligence approach to find the optimal relay node and power allocation. It initializes a population of particles that represents different solutions, and then iteratively updates their positions and velocities based on their fitness. The best particle is selected as the optimal solution. The basic steps of a PSO for CPD-NOMA can be summarized as follows [33]:

a. Initialize the swarm: Generate an initial swarm of particles, where each particle corresponds to a possible combination of power allocation, user clustering, and relay selection.

b. Evaluate fitness: Evaluate the fitness of each particle based on a fitness function that reflects the system performance in terms of sum rate and QoS requirements.

c. Update velocity and position: Update the velocity and position of each particle based on its current position, the position of the best-performing particle in the swarm, and the position of the best-performing particle that the particle has encountered so far.

d. Evaluation: Evaluate the fitness of the updated particles.

e. Update personal and global best: Update the personal best position of each particle based on its current position and the best position it has encountered so far. Update the global best position of the swarm based on the best position among all particles.

f. Termination: Repeat steps c to e until a termination condition is met, such as reaching a maximum number of iterations or achieving a desired level of performance.

g. The PSO can be used to optimize the power allocation, user clustering, and relay selection jointly to maximize the system performance in terms of sum rate while satisfying the QoS requirements of each user. The PSO can handle complex optimization problems and provide global optimality, but it may have high computational complexity and require a large number of iterations to converge to a solution.

iv. **Reinforcement learning (RL):** This algorithm uses a learning approach to find the optimal relay node and power allocation. It starts with an initial policy and iteratively improves it based on the reward function that reflects the system capacity. The optimal policy is learned through trial and error. The basic steps of an RL algorithm for CPD-NOMA can be summarized as follows [34]:

a. Define the state space: Define the state space of the system, which includes the channel conditions, the number of users, the available power, and other relevant parameters.

b. Define the action space: Define the action space of the system, which includes the power allocation, user clustering, and relay selection.

c. Define the reward function: Define the reward function of the system, which reflects the system performance in terms of sum rate and QoS requirements. The reward function can be designed to encourage the agent to maximize the sum rate while satisfying the QoS requirements of each user.

d. Train the agent: Train the agent using RL algorithms such as Q-learning, State-Action-Reward-State-Action (SARSA), or deep RL.

e. Evaluation: Evaluate the performance of the trained agent on a test set of scenarios that are similar to those encountered during training.

RL can be used to optimize the power allocation, user clustering, and relay selection jointly to maximize the system performance in terms of sum rate while satisfying the QoS requirements of each user. RL has the advantage of being able to adapt to changing environments and learn from experience, but it may have high computational complexity and require a large amount of training data.

Therefore, the joint power and relay selection algorithms for relay-assisted CPD-NOMA are important for achieving optimal performance in terms of capacity and power consumption. These algorithms can be adapted to different system requirements and constraints, such as the number of users, the channel conditions, and the power budget.

2.6 COMPARISON OF DF AND AF RELAY-ASSISTED CPD-NOMA

CPD-NOMA systems are designed to enhance the spectral efficiency and reliability of cellular communication networks. In addition to relay-assisted CPD-NOMA, there are other cooperative strategies that can be employed in CPD-NOMA systems. These strategies include amplify-and-forward (AF), decode-and-forward (DF), and hybrid cooperative strategies.

DF is a cooperative strategy in which the relay node decodes the message transmitted from the source and re-encodes it for transmission to the destination. This strategy is suitable for networks with strong relay-destination channels and high computational capabilities.

AF is a cooperative strategy in which the relay node amplifies and forwards the received signal from the source to the destination. This strategy is suitable for networks with weak relay-destination channels and low computational capabilities.

Hybrid cooperative strategies combine the advantages of DF and AF. In these strategies, the relay node first decodes the signal from the source and then amplifies and forwards it to the destination. This strategy is suitable for networks with moderate relay-destination channels and computational capabilities.

Compared to DF and AF, relay-assisted CPD-NOMA provides a higher spectral efficiency and can better cope with multi-user interference [35]. Moreover, it is less affected by the relay-destination channel quality and requires less computational complexity. However, the performance of relay-assisted CPD-NOMA is highly dependent on the power allocation strategy and channel conditions.

In conclusion, the selection of cooperative strategies in CPD-NOMA systems depends on the network characteristics and design requirements. Relay-assisted CPD-NOMA provides a promising solution for improving the spectral efficiency and reliability of wireless communication networks, but its performance should be compared with other cooperative strategies to determine the most suitable strategy for a specific network.

2.6.1 HYBRID COOPERATIVE STRATEGIES FOR CPD-NOMA SYSTEMS

Hybrid cooperative strategies of CPD-NOMA are a combination of DF and AF strategies [26]. In these strategies, the relay node first decodes the message received from the source and then amplifies and forwards it to the destination. This approach combines the advantages of DF and AF strategies and is suitable for networks with moderate relay-destination channels and computational capabilities.

In hybrid cooperative strategies, the relay node performs two main functions: decoding and forwarding. The decoding function is similar to DF, in which the relay node decodes the signal from the source and re-encodes it for transmission to the destination. The forwarding function is similar to AF, in which the relay node amplifies and forwards the received signal from the source to the destination.

The hybrid cooperative strategy can be divided into two stages: the first stage is the DF stage, in which the relay node decodes the signal from the source and re-encodes it for transmission to the destination, and the second stage is the AF stage, in which the relay node amplifies and forwards the signal to the destination.

The power allocation in hybrid cooperative strategies is a critical design parameter that affects the system performance. The power allocation scheme should be designed to maximize the sum rate of the system while satisfying the power constraints of the users and the relay node.

The advantages of hybrid cooperative strategies of CPD-NOMA are that they provide higher spectral efficiency than DF and AF strategies and are less affected by the relay-destination channel quality. Moreover, the computational complexity of hybrid cooperative strategies is lower than DF, making them more suitable for networks with moderate computational capabilities.

In conclusion, hybrid cooperative strategies of CPD-NOMA provide a promising solution for improving the spectral efficiency and reliability of wireless communication networks. The design of power allocation schemes and the selection of cooperative strategies depend on the network characteristics and design requirements.

2.6.2   INTEGRATION OF COOPERATIVE BEAMFORMING AND RELAY-ASSISTED CPD-NOMA SYSTEMS

Cooperative beamforming and relay-assisted CPD-NOMA [36] systems are two promising techniques to improve the spectral efficiency and reliability of wireless communication networks. The integration of these two techniques can provide even better performance gains by exploiting the benefits of both.

Cooperative beamforming is a technique that uses multiple antennas at the source and relay nodes to form a beam that focuses the signal toward the destination node. This technique can provide spatial diversity gain and mitigate the effects of fading and interference.

In relay-assisted CPD-NOMA systems, multiple users share the same time and frequency resources with the help of a relay node, and the power is allocated non-orthogonally to improve the system capacity. This technique can improve the spectral efficiency of the system and reduce the multi-user interference.

The integration of cooperative beamforming and relay-assisted CPD-NOMA systems can be achieved by using cooperative beamforming at the source and relay nodes and relay-assisted CPD-NOMA at the relay node. In this setup, the source and relay nodes use cooperative beamforming to form a beam toward the destination node, while the relay node uses relay-assisted CPD-NOMA to transmit the signals from the source node to the destination node.

The integration of cooperative beamforming and relay-assisted CPD-NOMA can provide several benefits. First, cooperative beamforming can enhance the reliability of the system by providing spatial diversity gain and reducing the fading effects. Secondly, relay-assisted CPD-NOMA can improve the spectral efficiency of the system by reducing the multi-user interference and allowing multiple users to share the same resources.

However, the integration of these two techniques also presents some challenges. The power allocation strategy needs to be carefully designed to optimize the system performance, taking into account the beamforming and CPD-NOMA constraints. Moreover, the design of the system needs to consider the channel conditions and user density to ensure efficient use of the available resources.

2.7   FUTURE RESEARCH DIRECTIONS

2.7.1   EMERGING TRENDS IN CPD-NOMA SYSTEMS

Here are some recent research papers that highlight the emerging trends in cooperative power domain NOMA:

i. **Machine learning (ML) and artificial intelligence (AI):** ML and AI techniques have the potential to optimize the performance of CPD-NOMA systems. These techniques can be used to optimize the power allocation, user clustering, and resource allocation. A comprehensive survey of the recent advances in ML-based resource allocation for cooperative NOMA networks.

ii. **Resource allocation:** Resource allocation is a critical issue in CPD-NOMA, and there are emerging trends toward developing efficient resource allocation strategies that can take into account the varying channel conditions, user densities, and power constraints. A joint power allocation and user clustering scheme for CPD-NOMA, which can improve the spectral efficiency of the system.

iii. **Energy harvesting (EH):** CPD-NOMA can also benefit from energy-harvesting technologies. EH techniques can be used to harvest energy from the environment and use it to power the devices in the system, which can reduce the dependence on external power sources. An energy-harvesting–enabled CPD-NOMA system can improve the energy efficiency and sustainability of the system.

iv. **Integration with other technologies:** CPD-NOMA can be integrated with other emerging technologies such as IoT, 5G, and mm-wave communications to create a more efficient and flexible system. The potential of CPD-NOMA for IoT applications highlights the emerging trends in this area.

v. **Security:** As CPD-NOMA becomes more prevalent, there is a growing need for security measures to protect the system from malicious attacks. Emerging trends in this area include the use of encryption and authentication techniques to secure the system. The security and privacy challenges in cooperative NOMA networks present potential solutions to address these challenges.

Therefore, the emerging trends in CPD-NOMA include the use of ML and AI, efficient resource allocation, EH, and integration with other technologies, security, and cognitive radio. In summary, CPD-NOMA is a rapidly evolving field, and there are several emerging trends that are shaping its development. These trends include the use of ML and AI, efficient resource allocation, EH, and integration with other technologies, security, and cognitive radio.

2.7.2 CHALLENGES IN CPD-NOMA SYSTEMS

Here are some recent research papers that highlight the challenges in cooperative power domain NOMA as shown in Figure 2.2.

A comprehensive survey of cooperative NOMA discusses the key challenges facing this technology, including power allocation, user clustering, and channel estimation. [37] proposes a joint power allocation and user clustering scheme for CPD-NOMA and highlights the challenges in implementing such a scheme, including the trade-off between system performance and computational complexity. [38] proposes an energy-harvesting–enabled CPD-NOMA system and discusses the challenges associated with integrating EH into NOMA, including the impact of EH on system performance and the complexity of EH-based resource allocation. [39]

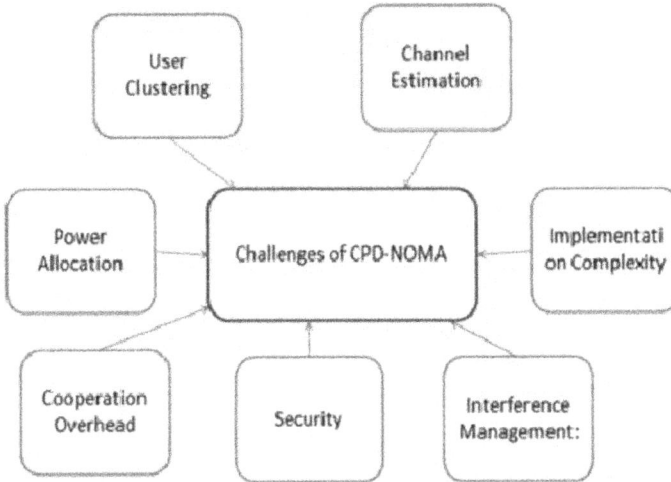

FIGURE 2.2 Challenges of CPD-NOMA.

discusses the security and privacy challenges in cooperative NOMA networks, including eavesdropping, interception, and unauthorized access. Therefore, these papers highlight the challenges facing CPD-NOMA, including power allocation, user clustering, channel estimation, EH, security and privacy, and integration with other technologies.

While CPD-NOMA has the potential to improve the spectral efficiency and energy efficiency of wireless networks, there are several challenges that need to be addressed. Here are some of the key challenges facing CPD-NOMA:

  i. **Power allocation:** Power allocation is a critical issue in CPD-NOMA, and it can be challenging to allocate power efficiently among the users while maintaining a low level of interference.
  ii. **User clustering:** User clustering is another key challenge in CPD-NOMA. Clustering users with similar channel conditions can improve the system performance, but it can be challenging to identify and cluster users efficiently in a dynamic environment.
  iii. **Channel estimation:** Accurate channel estimation is essential for CPD-NOMA to work effectively. However, channel estimation can be challenging in a multi-user environment with varying channel conditions and interference.
  iv. **Implementation complexity:** The implementation complexity of CPD-NOMA can be high, particularly in large-scale systems. The hardware and software required to implement cooperative PD-NOMA can be expensive and complex.
  v. **Interference management:** Interference is a significant challenge in CPD-NOMA, particularly when multiple users are transmitting at the same time. Efficient interference management techniques are required to reduce interference and improve system performance.

vi. **Security:** As with any wireless communication system, security is a significant concern for CPD-NOMA. Security threats, such as eavesdropping and hacking, can compromise the integrity of the system and jeopardize user privacy.

vii. **Cooperation overhead:** CPD-NOMA requires cooperation among the users, which can add additional overhead to the system. The overhead can increase as the number of users in the system grows, which can impact the overall system performance.

In summary, CPD-NOMA faces several challenges that need to be addressed for the technology to reach its full potential. These challenges include power allocation, user clustering, channel estimation, implementation complexity, interference management, security, and cooperation overhead.

2.8 CONCLUSION

This study offered a thorough review of CPD-NOMA for 5G wireless networks. According to the studies mentioned, research in this area is still in its initial stages, thereby leaving the subject open for future research. In emerging technologies, integrated networks must meet stated objectives such as efficiency, reliability, and greater system performance. The performance and advantages of NOMA over OMA are also discussed in this chapter. In addition, different types of cooperative strategies for CPD-NOMA and their advantages are discussed. Finally, open research challenges are discussed.

REFERENCES

1. Ghafoor, U., Ali, M., Khan, H. Z., Siddiqui, A. M., & Naeem, M. (2022). NOMA and future 5G & B5G wireless networks: A paradigm. Journal of Network and Computer Applications, 204, 103413.
2. Goyal, J., Singla, K., & Singh, S. (2020). A survey of wireless communication technologies from 1G to 5G. In Second International Conference on Computer Networks and Communication Technologies: ICCNCT 2019 (pp. 613–624). Springer International Publishing.
3. Huang, M. Y., Chen, Y. W., Shiu, R. K., Wang, H., & Chang, G. K. (2021). A bi-directional multi-band, multi-beam mm-wave beamformer for 5G fiber wireless access networks. Journal of Lightwave Technology, 39(4), 1116–1124.
4. AI_Dujaili, M. J., & Salih, B. A. (2021). A review of mobile technologies from 1G to the 5G and a comparison between them. Solid State Technology, 64(2), 2805–2823.
5. Scott, A. W., & Frobenius, R. (2008). Multiple access techniques: FDMA, TDMA, and CDMA.
6. Bikos, A. N., & Sklavos, N. (2012). LTE/SAE security issues on 4G wireless networks. IEEE Security & Privacy, 11(2), 55–62.
7. Li, H., Ru, G., Kim, Y., & Liu, H. (2010). OFDMA capacity analysis in MIMO channels. IEEE Transactions on Information Theory, 56(9), 4438–4446.
8. Dangi, R., Lalwani, P., Choudhary, G., You, I., & Pau, G. (2022). Study and investigation on 5G technology: A systematic review. Sensors, 22(1), 26.

9. Makki, B., Chitti, K., Behravan, A., & Alouini, M. S. (2020). A survey of NOMA: Current status and open research challenges. IEEE Open Journal of the Communications Society, 1, 179–189.
10. Reddy, P. V., Reddy, S., Reddy, S., Sawale, R. D., Narendar, P., Duggineni, C., & Valiveti, H. B. (2021, October). Analytical review on OMA vs. NOMA and challenges implementing NOMA. In 2021 2nd International Conference on Smart Electronics and Communication (ICOSEC) (pp. 552–556). IEEE.
11. Ganji, M., & Jafarkhani, H. (2019, April). Time asynchronous NOMA for downlink transmission. In 2019 IEEE Wireless Communications and Networking Conference (WCNC) (pp. 1–6). IEEE.
12. Hao, X., Wu, Y., Xiao, Y., Chen, Z., & Atkin, G. E. (2021, September). Time offsets format for NOMA system. In 2021 IEEE 94th Vehicular Technology Conference (VTC2021-Fall) (pp. 1–6). IEEE.
13. Wang, R., Thompson, J. S., & Haas, H. (2010, March). A novel time-domain sleep mode design for energy-efficient LTE. In 2010 4th International Symposium on Communications, Control and Signal Processing (ISCCSP) (pp. 1–4). IEEE.
14. Maraqa, O., Rajasekaran, A. S., Al-Ahmadi, S., Yanikomeroglu, H., & Sait, S. M. (2020). A survey of rate-optimal power domain NOMA with enabling technologies of future wireless networks. IEEE Communications Surveys & Tutorials, 22(4), 2192–2235.
15. Islam, S. M., Zeng, M., & Dobre, O. A. (2017). NOMA in 5G systems: Exciting possibilities for enhancing spectral efficiency. arXiv preprint arXiv:1706.08215.
16. Thirumavalavan, V. C., & Jayaraman, T. S. (2020, January). BER analysis of reconfigurable intelligent surface assisted downlink power domain NOMA system. In 2020 International Conference on Communication Systems & Networks (COMSNETS) (pp. 519–522). IEEE.
17. Yue, X., Qin, Z., Liu, Y., Kang, S., & Chen, Y. (2018). A unified framework for non-orthogonal multiple access. IEEE Transactions on Communications, 66(11), 5346–5359.
18. Khan, W. U., Lagunas, E., Mahmood, A., Ali, Z., Chatzinotas, S., & Ottersten, B. (2022). Integration of NOMA with reflecting intelligent surfaces: A multi-cell optimization with SIC decoding errors. arXiv preprint arXiv:2205.03248.
19. Senel, K., Cheng, H. V., Björnson, E., & Larsson, E. G. (2019). What role can NOMA play in massive MIMO? IEEE Journal of Selected Topics in Signal Processing, 13(3), 597–611.
20. Ding, Z., Fan, P., & Poor, H. V. (2017). Random beamforming in millimeter-wave NOMA networks. IEEE access, 5, 7667–7681.
21. Salem, A., & Musavian, L. (2020). NOMA in cooperative communication systems with energy-harvesting nodes and wireless secure transmission. IEEE Transactions on Wireless Communications, 20(2), 1023–1037.
22. Lee, S. R., Park, S. H., & Lee, I. (2018). NOMA systems with content-centric multicast transmission for C-RAN. IEEE Wireless Communications Letters, 7(5), 828–831.
23. Ghous, M., Hassan, A. K., Abbas, Z. H., Abbas, G., Hussien, A., & Baker, T. (2022). Cooperative power-domain NOMA systems: An overview. Sensors, 22(24), 9652.
24. Lim, G., & Cimini, L. J. (2013). Energy-efficient cooperative beamforming in clustered wireless networks. IEEE Transactions on Wireless Communications, 12(3), 1376–1385.
25. Ramadan, A., Zorba, N., & Hassanein, H. S. (2022, December). Uplink cluster-based radio resource scheduling for HetNet mMTC scenarios. In GLOBECOM 2022-2022 IEEE Global Communications Conference (pp. 1–6). IEEE.
26. Ramadan, A. M. (2022). Uplink cluster-based radio resource scheduling and allocation for HetNet mMTC scenarios (Doctoral dissertation, Queen's University (Canada)).

27. Gure, H. (2021). Performance comparison of multiple access techniques for 5G networks and beyond (Doctoral dissertation).

28. Chen, H., Zhai, C., Li, Y., & Vucetic, B. (2018). Cooperative strategies for wireless-powered communications: An overview. IEEE Wireless Communications, 25(4), 112–119.

29. Cox, H., Zeskind, R., & Owen, M. (1987). Robust adaptive beamforming. IEEE Transactions on Acoustics, Speech, and Signal Processing, 35(10), 1365–1376.

30. Li, G., Mishra, D., Hu, Y., & Atapattu, S. (2020). Optimal designs for relay-assisted NOMA networks with hybrid SWIPT scheme. IEEE Transactions on Communications, 68(6), 3588–3601.

31. Yu, N. Y. (2019). Multiuser activity and data detection via sparsity-blind greedy recovery for uplink grant-free NOMA. IEEE Communications Letters, 23(11), 2082–2085.

32. Cetinkaya, S., & Arslan, H. (2020, June). Energy and spectral efficiency tradeoff in NOMA: Multi-objective evolutionary approaches. In 2020 IEEE International Conference on Communications Workshops (ICC Workshops) (pp. 1–6). IEEE.

33. Hao, S. W., Li, Y. J., Zhao, S. H., Wang, W. L., & Wang, X. Y. (2020). Multicarrier NOMA power allocation strategy based on improved particle swarm optimization algorithm. Acta Electonica Sinica, 48(10), 2009.

34. Rauniyar, A., Yazidi, A., Engelstad, P., & Østerbo, O. N. (2020, November). A reinforcement learning based game theoretic approach for distributed power control in downlink NOMA. In 2020 IEEE 19th International Symposium on Network Computing and Applications (NCA) (pp. 1–10). IEEE.

35. Naeem, M., Anpalagan, A., Jaseemuddin, M., & Lee, D. C. (2013). Resource allocation techniques in cooperative cognitive radio networks. IEEE Communications Surveys & Tutorials, 16(2), 729–744.

36. Song, N., Yang, T., & Sun, H. (2019, April). Efficient hybrid beamforming for relay assisted millimeter-wave multi-user massive MIMO. In 2019 IEEE Wireless Communications and Networking Conference (WCNC) (pp. 1–6). IEEE.

37. Shili, M., Hajjaj, M., & Ammari, M. L. (2022). User clustering and power allocation for massive MIMO with NOMA-inspired cognitive radio. IEEE Transactions on Vehicular Technology, 71(7), 7656–7664.

38. Guo, C., Zhao, L., Feng, C., Ding, Z., & Chen, H. H. (2019). Energy harvesting enabled NOMA systems with full-duplex relaying. IEEE Transactions on Vehicular Technology, 68(7), 7179–7183.

39. Wei, L., Chen, Y., Zheng, D., & Jiao, B. (2020). Secure performance analysis and optimization for FD-NOMA vehicular communications. China Communications, 17(11), 29–41.

3 6G Ultra-Dense O-RAN Heterogeneous Network and Its Performance Analysis

Gopal Chandra Das, Abhishek Kumar,
Seemanti Saha, and Abhijit Bhowmick

3.1 INTRODUCTION

The connectivity crisis of customers is the present problem in automation technology for business units, security, activities, developers, and infrastructure that
offers accessible, effective, and agent-less applications. With the help of customer-
generated needs, it is possible to measure the effectiveness of the network and
resource utilization for smart projects in smart cities [1]. Future 6G communication will require deep and intelligent, holographic, and pervasive connectivity. Future 6G network design is becoming more heterogeneous and vast and has
various goals with self-optimization to serve different, dynamic, and demanding
business models with AI approaches [2]. Moreover, zero-touch service administration, automation, security, and trust networking capabilities are all included in
network softwarization, which are equally crucial [3]. The 6G system integrates
several technologies with intelligence capabilities, connecting several emerging
technologies. The combination of these dynamically versatile evolving technologies, which would assist future generations in meeting the complex user demands
of the development of smart projects, presents a challenge for the network of the
future generation [4]. Hence, 6G should be the requirement that 5G cannot meet
and needs further improvement to meet society's information needs [5]. The peak
transmission rate of the 6G network will be 10 Gbps, density of the traffic will be
10 TBPs/km^2, the connection density will be 1 million sets/km^2, and the user experience rate will be 0.1–1Gbps, according to the 6G standard proposal [6]. End-to-end
latency reaches the ms level at 500 km/h, which can guarantee the user experience.
Moreover, it dramatically improves energy, cost, and spectrum efficiency. Future
scenarios for using the 5G network will increase, and numerous experts in the field
are already working on the research and development of 6G network applications
for unmanned aerial vehicles (UAVs), logistics, vehicle networking, etc. Radio
frequency spectrum management has grown in importance in today's expanding
globe concerning wireless communications [7]. Radio frequency (RF) spectrum
scarcity is the main challenge facing developing wireless communication network

systems [8]. Several research and development initiatives are already tackling 6G mobile systems, mostly focusing on emerging and translating technologies, even if fifth-generation (5G) systems have not yet been deployed in several nations [9]. To succeed commercially, it will be necessary to scale publicly accessible technologies like virtualization, AI, and cloud radio access networks [10]. To simplify and hasten the adoption of multi-vendor radio access networks (RANs) in 6G mobile networks, the Open Radio Access Network (O-RAN) Alliance is a step in the right direction [11]. The O-RAN architectures for 6G mobile are currently being investigated as a heterogeneous network [12]. This technology is a strong contender for next-generation mobile networks since it can handle enormous capacities and numerous concurrent users. This technique can take care of fully integrated and extremely dense heterogeneous network solutions that profit from software stations and commoditized hardware. Optimizing 6G network resources using O-RAN has recently been the subject of extensive research [13]. An intelligent radio resource management strategy is proven to be useful for reducing traffic congestion, and the plan's efficacy is illustrated using real data from a significant operator [14]. In [15], the author discusses the situations that inspired the idea of 6G, reveals their desired 6G performance requirements, explains the supporting technologies for the 6G services they have already introduced, and proposes a new set of service classes. The authors describe the two core technologies of computational holographic radio and photonics-based cognitive radio, which are predicted to become the key 6G technologies necessary for these services [16]. They claim these services will promote the merger of photonics and AI and necessitate an end-to-end processing, sensing, and communication code. Using heuristic techniques, [17] suggests network slicing and joint power allocation for an O-RAN system with challenging sub-problems of high computational complexity. The 5G slice-in-slice scheduling optimization outcomes are shown in [18]. Joint resource allocation and UD selection in O-RAN with guaranteed energy efficiency and latency are presented in [19]. [20] presents radio interference and resource management (RIRM) strategies to address the difficulties in efficiently using the resources already at hand, including spectrum band, energy, modulation scheme, cyclic prefix, and guard band allocation.

Li and Qu [21] suggested multi-objective particle swarm optimization (MOPSO) in 2020. The multi-objective version of particle swarm optimization (PSO) uses the Pareto envelope and the four lattice generation technique to solve multi-objective optimization issues. In contrast to PSO, more than one criterion is used to determine and define the most (local or global) optimal solution. Dominance-based and probabilistic principles are applied for the personal (local) best particle. It is anticipated that resource allocation will be heavily researched in future years at small-base stations (SBS) and macro-cell base stations (MBS) regarding local and global maximum capabilities. The AI-based resource optimization in the O-RAN network is the main focus of the literature mentioned above. However, to the authors' knowledge, the ultra-dense heterogeneous O-RAN cellular system design has yet to be published. As a result, rather than using AI-based optimization in ultra-dense heterogeneous scenarios, we are considering MOPSO at both SBSs and MBSs in ultra-dense O-RAN cellular systems. This will enable us to identify the enabling technologies for the 6G services that are being proposed and to disclose their intended 6G performance requirements.

3.2 SYSTEM MODEL DESCRIPTION

By offering a non-proprietary version of the RAN system, the O-RAN is a RAN that enables cellular network equipment from several suppliers to work together. With integrating new technologies into cellular networks today, such as massive multiple-input multiple-output (MIMO), millimeter wave, and sub-terahertz communications, network operators face the difficult task of managing the rising complexity and costs of maintaining these systems to keep up with consumer demands and market trends. Higher data speeds, lower latency, more connection, and innovative applications and services offered by the 6G network are anticipated to transform the telecommunications sector completely. However, cooperation across various networks and technologies, such as 5G, wi-fi, satellite, and other wireless technologies, is necessary to realize the goal of a 6G network.

Figures 3.1 and 3.2 depict the deployment scenario structure and end-to-end flow of the suggested solution for the O-RAN architecture. Figure 3.1 shows a schematic block diagram of the distributed O-RAN architecture. The four fundamental building elements of O-RAN are the physical layer, the network layer, the application interface, and the network control and service management system. Several crucial elements enable the O-RAN architecture to function with heterogeneous networks to build a 6G network. The RAN intelligent controller (RIC), the central unit (CU), the distributed unit (DU), and the radio unit (RU) are some of these parts. An essential element of the O-RAN architecture that makes it possible for the network to be programmed and automated is the RIC. RAN functions, such as managing radio resources, mobility, and quality of service (QoS), are under the authority and management of the RIC. The RIC provides a standardized interface for various network components, enabling seamless integration and interoperability. Radio resource management, mobility management, and QoS management are all RAN functions that must be controlled and managed by the CU. The CU offers a centralized

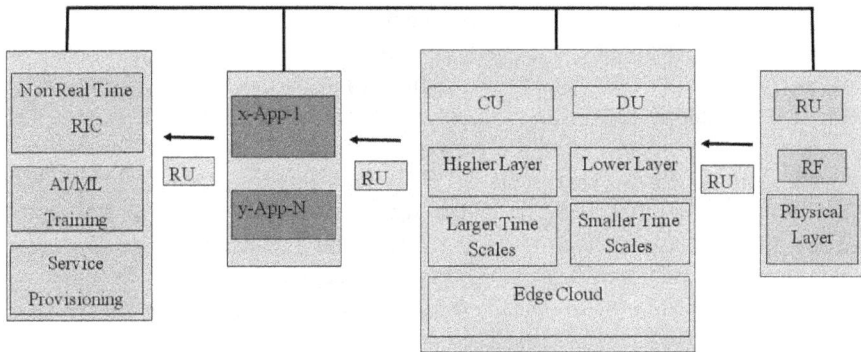

FIGURE 3.1 A distributed O-RAN architecture for 6G communication.

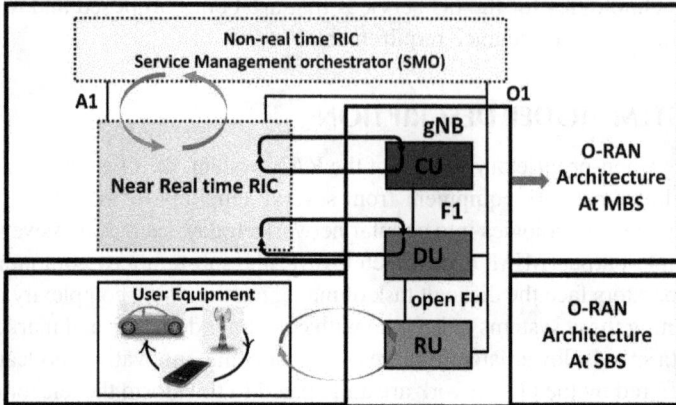

FIGURE 3.2 System model for resource allocation based on O-RAN heterogeneous network.

abstraction layer to manage and optimize the network's components. The CU can also employ real-time analytics and machine-learning strategies to improve network performance. The network's physical layer operations, such as the modulation and demodulation of the radio signals, are handled by the DU. The DU offers a distributed abstraction layer that can manage and optimize the network's physical layer components. The RU sends and receives radio signals between the user equipment and the network. The RU provides a standardized interface that can integrate different radio technologies, including 5G, wi-fi, and other wireless technologies.

An RU and a DU are responsible for seamless interfacing with the heterogeneous user equipment that is part of low-power SBSs in an ultra-dense 6G cellular network architecture, as shown in Figure 3.3. A central unit (CU) is part of the MBSs connected to multiple DUs to provide connectivity to the core mobile and data network. As shown in Figure 3.2, the service management and orchestrator (SMO), non-real-time intelligent controller (non-RT-RIC), and Near RIC are the O-RAN entities that supervise the interactions between RU, DU, and CU for efficient and optimum resource utilization. The near-RT RIC and the non-RT-RIC are connected through the A1 interface, one of the O-RAN-specific interfaces that enables the deployment of intelligent models, policy, and guidance in the near-RT RIC's non-real-time control loop. The non-RT RIC also terminates the O1 interface, which connects to every other RAN component and is used to control and coordinate network activity. The task is divided or controlled by F1 between CU and DU. To locate the intra-cell network's call network, open fronthaul (FH) is used. The non-real-time (or non-RT) RIC is a component that works in tandem with the near-real-time (RT) RIC in the SMO framework, as depicted in Figure 3.2, to enable intelligent O-RAN operation and optimization on a time scale longer than one second. The non-RT RIC uses the non-real-time control loop to manage machine learning (ML) models for the near-RT RIC and to give advice and enrichment data. Additionally, the non-RT RIC has indirect control over all O-RAN architectural elements linked to the SMO due to its ability to influence SMO activities. The near-RT closed-loop control may have an

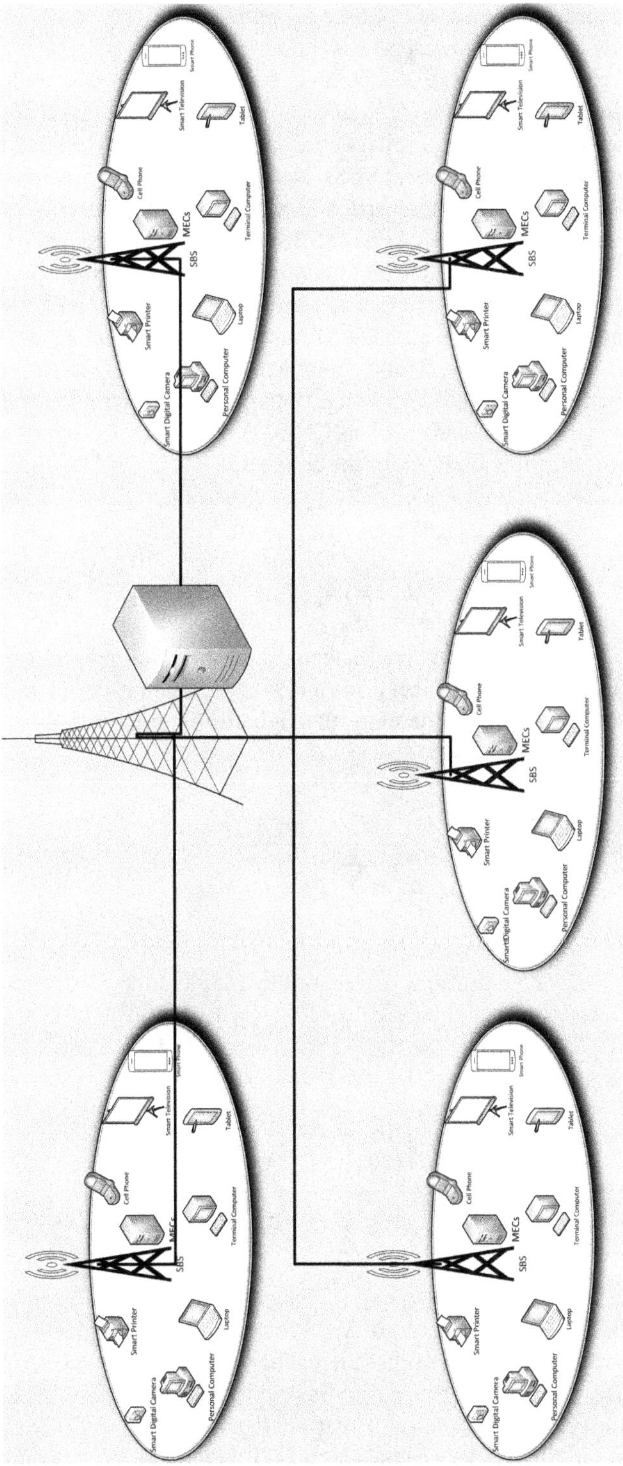

FIGURE 3.3 Multi-tier energy allocation in ultra-dense 6G heterogeneous cellular network architecture.

influence on hundreds or thousands of user equipments (UEs) because the near-RT-RIC is frequently connected to several RAN nodes.

An O-RAN-based ultra-dense 6G heterogeneous cellular network architecture is considered in this work and shown in Figure 3.3. Depending on the cell architecture, such as femtocells, picocells, and microcells, each tiny cell is estimated to include 100–500 UE connected by low-power SBSs. Next, an optical link connects each SBS to a homogeneous cell MBS. These MBSs serve as a hub and connect users to the leading mobile and data network. At both SBSs and MBSs, we suggest simultaneously optimizing energy and spectrum allocation for fading channels with delayed CSI. Assume that X UEs are linked to SBSs and that a group of Y SBSs is linked to MBSs. The channel gain $g_{x,y}$ between the UE and the Yth SBS and $g_{x,y} = |h_{x,y}|^2 \alpha_{x,y}$ where $h_{x,y}$ is the small-scale fast fading component and $\alpha_{x,y}$ is the large-scale fading effects factor such path loss and shadowing. Both pieces are assumed to be independently and identically distributed (i.i.d.) as $CN(0, 1)$. The SBSs are expected to know the channel status information (CSI) of the connected UE to the SBS link. Next, the variation in channel over time T is modeled by the first-order Gauss-Markov process, given by

$$h_{x,y} = \varepsilon \hat{h}_{x,y} + e \tag{3.1}$$

here, $\hat{h}_{x,y}$ and $h_{x,y}$ are the channel coefficients in the past and present time respectively, $e \in CN(0, 1-\varepsilon^2)$ is the channel error and ε is the channel correlation between the two successive time slots. Therefore the signal-to-interference-plus-noise-ratio (SINR) of the receiver at SBS is

$$\gamma_{SBS} = \frac{P_x \alpha_{x,y} |h_{x,y}|^2}{\sigma_0^2 + \sum\limits_{j=1}^{X} \rho_{j,x} P_j \alpha_j |\hat{h}_{j,y}|^2}. \tag{3.2}$$

Where P_x and P_j denote transmit powers of the xth E and the jth UE, respectively, σ_0^2 is received noise power, and maybe 1 or 0, $\rho_{j,x} = 1$ when j^{th} UE reuses the spectrum of the xth UE else $\rho_{j,x} = 0$. Next, Y^{th} SBSs send their data to MBS. Further, the SINR at MBS is given by

$$\gamma_{MBS} = \frac{\left(P_y \alpha_y \left(\varepsilon_y^2 |\hat{h}_y| + |\hat{e}_y|^2\right)\right)}{\sigma^2 + \sum\limits_{y=1}^{Y} \rho_{y'} P_{y'} \alpha_y |\hat{h}_{y'}|^2}. \tag{3.3}$$

Here, $\hat{e}_y$ is the AWGN noise vector at MBS received signal. To meet the high data requirement for 5G cellular networks, the capacity for SBSs connection with UE and MBS connectivity to the SBSs has to be optimized for highly reliable resource allocation and to get maximum channel capacity between UE to SBSs and SBS to MBS. Therefore, in this work, we consider a MOPSO algorithm to simultaneously solve the channel capacity optimization of both SBSs and MBSs. Further, the capacity maximization problem is given by

$$\max_{\rho_{j,x},P_x,P_j} \sum_{x \in X} log_2(1+\gamma_{SBS}), \tag{3.4}$$

$$log_2(1+\gamma_{SBS}) \geq r_x, \qquad \forall x \in X, \tag{3.5}$$

$$\sum_{x \in X} \rho_{j,x} \leq 1, \qquad \rho_{j,x} \in \{0,1\} \qquad \forall x \in X, \tag{3.6}$$

and

$$\max_{\rho_{y'},P_y,P_{y'}} \sum_{y \in Y} log_2(1+\gamma_{MBS}), \tag{3.7}$$

$$log_2(1+\gamma_{MBS}) \geq r_0, \qquad \forall y \in Y, \tag{3.8}$$

and

$$\sum_{y \in Y} \rho_{y'} \leq 1, \qquad \rho_y \in \{0,1\} \qquad \forall y \in Y, \tag{3.9}$$

where r_x is the minimum acceptable transfer rate for X^{th} UE and r_0 is the minimum reasonable transfer rate for Y^{th} MBS.

3.3 PROPOSED RESOURCE ALLOCATION DESIGN

Power allocation for MBS to SBSs and SBS to UEs communication links can be expressed as bellow.

The channel capacity for MBS to X^{th} SBS and SBS to UEs are given by

$$C_{x,y} = log_2 \left(1 + \frac{P_x^c \alpha_{x,c} |h_{x,c}|^2}{\sigma^2 + P_y^d \alpha_y |\hat{h}_y|} \right). \tag{3.10}$$

While satisfying all the constraints, the optimization problem can be formulated by

$$\max_{P_y,P_x} C_{x,y}. \tag{3.11}$$

A feasible region from (11) is deduced as [18]

$$\left\{ \begin{array}{l} \left(P_y^d, P_x^c\right) : \exp\left(\frac{R\gamma_0^d}{Q}\right) \left(1 + \frac{S}{Q}\gamma_0^d\right) \leq \dfrac{\exp\left(\dfrac{P}{Q}\right)}{1-\rho_0}, R\gamma_0^d \geq P, 0 \leq P_x^c \leq P_{max}^c, \\ \\ 0 \leq P_y^d \leq P_{max}^d \end{array} \right\} \tag{3.12}$$

or

$$\left\{ \begin{array}{l} \left(P_y^d, P_x^c\right): \left(1 + \dfrac{Q}{\gamma_0^d D}\right) \exp\left(\dfrac{P - R\gamma_0^d}{\gamma_0^d S}\right) \geq \dfrac{1}{\rho_0}, \quad C\gamma_0^d \langle A, 0 \leq P_x^c \leq P_{\max}^c, \\ \\ 0 \leq P_y^d \leq P_{\max}^d \end{array} \right\}, \quad (3.13)$$

where

$$P = p_y^d \alpha_y \varepsilon_y^2 \left| \hat{h}_y \right|, \quad (3.14)$$

$$Q = p_y^d \alpha_y \left(1 - \varepsilon_y^2\right), \quad (3.15)$$

$$R = \sigma^2 + p_x^c \alpha_{x,y} \left| \hat{h}_{x,y} \right|^2, \quad (3.16)$$

and

$$S = p_x^c \alpha_{x,y} \left(1 - \varepsilon_{x,y}^2\right). \quad (3.17)$$

The proof of (13) is presented in [18] by which the optimal power allocation at SBS and MBS is derived as follows:

$$p_y^{d*} = \left\{ \begin{array}{l} \min\left\{\left(p_{\max}^d, p_{c,\max}^{d1}\right)\right\} if p_{\max} \leq p_0^d \\ \min\left\{\left(p_{\max}^d, p_{c,\max}^{d2}\right)\right\} if p_{\max}^d \geq p_0^d, p_{\max}^c \rangle p_0^c \\ p_{c,\max}^{d1} \end{array} \right. \quad (3.18)$$

and

$$p_x^{c*} = \left\{ \begin{array}{l} \min\left\{\left(p_{\max}^c, p_{d,\max}^{c1}\right)\right\} if p^d \leq p_0^d \\ \min\left\{\left(p_{\max}^c, p_{d,\max}^{c2}\right)\right\} if p_{\max} \geq p_0^d, p_{\max}^c \rangle p_0^c \\ p_x ax^{c1} \end{array} \right. , \quad (3.19)$$

where

$$p_0^c = \dfrac{\sigma^2}{\dfrac{1 - \varepsilon_{x,y}^2}{1 - \varepsilon_y^2}\left(\dfrac{1}{\rho_0} - 1\right)\alpha_y, k\varepsilon_y^2 \left|\hat{h}_y\right|^2 - \alpha_x, k\varepsilon_{x,y}^2 \left|\hat{h}_{x,y}\right|^2} \quad (3.20)$$

and

$$p_0^d = \dfrac{p_0^c \gamma_0^d \alpha_{x,y} k(1 - \varepsilon_{x,y}^2)(1 - \rho_0^2)}{\alpha_y (1 - \varepsilon_y^2)\rho_0^2}, \quad (3.21)$$

where $p_{c,\max}^{d1}$ and $p_{d,\max}^{c1}$ are originated with the implicit functions $F_1\left(P_{c,\max}^{d1}, P_{\max}^{c}\right) = 0$ and $F_1\left(P_{\max}^{d}, P_{d,\max}^{c1}\right) = 0$ as well as $p_{c,\max}^{d2}$ and $p_{d,\max}^{c2}$ are designed with the implicit functions $F_2\left(P_{\max}^{d}, P_{d,\max}^{c2}\right) = 0$ and $F_2\left(P_{c,\max}^{d2}, P_{\max}^{c}\right) = 0$. Using the bisection search to note the monotonic relationship between p_x^c and p_y^d in the implicit functions

$$F_1\left(P_y^d, P_x^c\right) = \left(1 + \frac{S}{Q}\gamma_0^d\right)\exp\left(\frac{R\gamma_0^d}{Q}\right) - \frac{\exp\left(\dfrac{P}{Q}\right)}{1 - \rho_0} = 0, \tag{3.22}$$

where $p_y^d \in \left\{0, p_0^d\right\}$ and

$$F_2\left(P_y^d, P_x^c\right) = \left(1 + \frac{Q}{\gamma_0^d S}\right)\exp\left(\frac{P - \gamma_0^d}{\gamma_0^d S}\right) - \frac{1}{\rho_0} = 0 \tag{3.23}$$

where $p_y^d \in \left\{p_0^d + \infty\right\}$.

3.4 SIMULATION RESULT

This section presents a complete simulation analysis to validate the suggested approach. We have considered $f_c = 2.5$ GHz and 5 GHz, SNR from 0–30 dB. The total power availability for SBSs and MBSs links is from 0–20 dB. Figure 3.3 depicts the system model used to simulate the proposed process. The recommended method and the ORAN-based network architecture are correlated in Figure 3.2. The heterogeneous UEs' SBS zone is thought to have a radius of 100–500 m². Figures 3.4 to 3.6 show the simulation study of the proposed MOPSO-based resource allocation for 6G ultra-dense heterogeneous networks. In an O-RAN-based ultra-dense heterogeneous network, the MOPSO function's feasibility region is simulated using the maximum

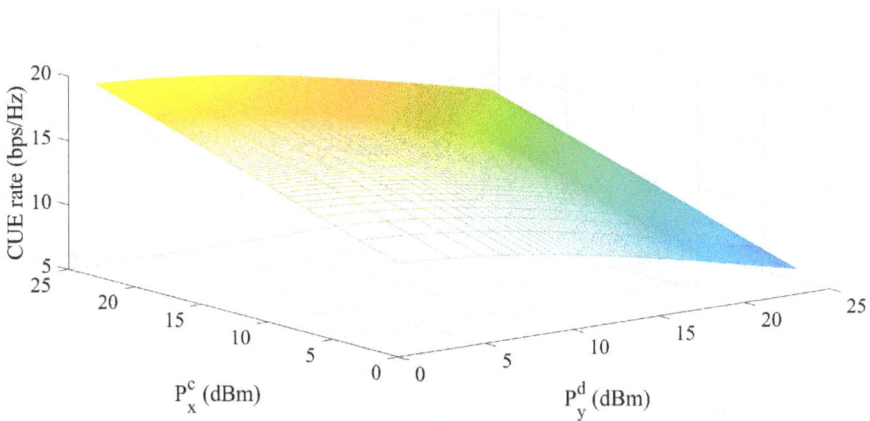

FIGURE 3.4 A 3D plot for the feasible region of optimal power allocation for both MBS and SBS.

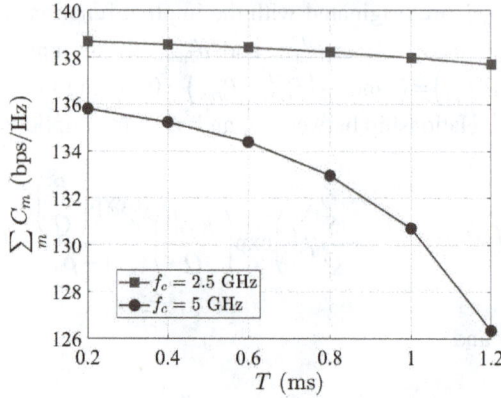

FIGURE 3.5 A normalized, optimized sum capacity rate plot for the proposed technique.

power allotted for the links from SBS to UEs and MBSs to SBSs. Figure 3.4 shows a 3D graphic for the functional area of optimal power allocation for both MBS and SBS and the corresponding UE bit rate capacity. It has been observed that the maximum UE rate capacity can be achieved when the power is approximately equally distributed among UEs' and SBSs' links.

Figure 3.5 shows the likelihood of a UE and SBS link outage using the MOPSO algorithm. UEs SINR is kept at 5 dB to examine the CDF of outage probability. Higher interference power scenarios have been found to increase the likelihood of outages. In contrast, higher carrier frequencies improve the performance of the proposed algorithm because they narrow the spectrum beam width and reduce interference power.

Figure 3.6 shows the cumulative UE rate capacity linked to preserving 5dB noise variance at SU for various carrier frequencies. The figure shows that the rate is

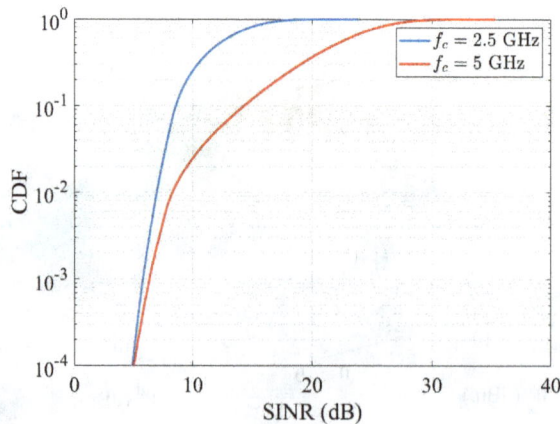

FIGURE 3.6 A cumulative outage link probability between the UEs and SBSs for the proposed technique.

constant for $f_c = 2.5$ GHz and gradually declines for higher carrier frequencies. This is because the proposed scheme is designed to serve more UEs in a given area at higher frequencies, which adds to the scheme's complexity and causes a slow decline in rate performance due to increased UE-related interference.

3.5 CONCLUSION

Instead of an AI-based resource optimization in this paper's high-density hetero-geneous scenario due to the heterogeneity of the multiple UEs, we have provided MOPSO at both SBS and MBS in a high-density O-RAN cellular system. The exper-imental results demonstrate not only that the suggested method has a discernible impact on the optimization of the system's energy consumption in many circum-stances, but it also demonstrates that this algorithm has a high degree of conver-gence. The simulation analysis demonstrates that the suggested algorithm ensures the minimal UE SINR power of 5dB and achieves the sum capacity rate by minimiz-ing the failure probability link.

AUTHORS' BIOGRAPHICAL DETAILS

Gopal Chandra Das received the B.Sc Engineering degree in Electronics Engineering from the National Institute of Technology, Jamshedpur, India, in 2003. He has obtained his M.Tech degree in Electronics and Telecommunication from the National Institute of Technology, Durgapur, India in 2009. Currently he is pursuing his Ph.D from the Department of Electronics and Communication Engineering, NIT Patna, Patna, India and also is working as an Assistant Professor in the department of Electronics and Communication at Bengal College of Engineering and Technology, Durgapur, India. His research interests include Cognitive Radio Networks focusing on Spectrum Sensing and Spectrum Sharing issues, Cooperative Communications in Cognitive Radio Networks, energy harvesting in wireless network and D2D communication. He has published many research papers in various journals and conferences.

Abhishek Kumar received the B.Tech. degree in Electronics and Telecommunication Engineering from IETE, New Delhi, India, in 2011 and the M. Tech in Digital Communication from USICT Guru Govind Sing Indraprastha University, New Delhi, India, in 2014. He has obtained his Ph.D. from the Department of Electronics and Communication Engineering, NIT Patna, Patna, India. He is currently working as a post-doctoral fellow at CentraleSupelec, Paris Saclay University, Rennes, France. Previously, he worked as a

Research Associate at IIT Kharagpur in MeitY sponsored 5G and Beyond Project. He worked as Assistant Professor at BIET Hyderabad from 2020–2021. He has published many research papers in various journals and conferences.

Seemanti Saha received the B.Tech. degree in electronics and communication engineering from the University of Kalyani, India, in 2002 and the M.Tech. degree in electronics and electrical communication engineering with specialization in telecommunication system engineering from IIT Kharagpur, India, in 2006. She has obtained her Ph.D. degree from IIT Kharagpur, India, in 2014 in the area of wireless communication. Presently she is working as an assistant professor at the National Institute of Technology Patna, India since 2013. Her primary research interests include detection and estimation in wireless communication systems, physical layer algorithms in communication engineering, statistical and adaptive signal processing for telecommunications, machine learning and its application in communication engineering, fuzzy systems and models, and the application of fuzzy systems in communication engineering. She serves as a reviewer of internationally reputed journals such as IEEE, Elsevier, Springer, etc.

Abhijit Bhowmick received his B.E (Hons) degree in Electronics and Telecommunication Engineering in 2002 from Burdwan University, West Bengal, India, M.Tech. degree in Telecommunication Engineering in 2009 and his Ph.D. in 2016 from NIT Durgapur. He worked for Cubix Control System Pvt. Ltd. from 2004–2006. He joined in the Department of Electronics and Comm. Engineering, Bengal College of Eng'g. and Tech., Durgapur as a lecturer in 2006. Thereafter, he joined VIT University, Vellore, TN, India in the School of Electronics Engineering (SENSE) in June 2016 and is currently serving as an Associate Professor there. His research interests include Cognitive Radio Networks focusing on Spectrum Sensing and Spectrum Sharing issues, Cooperative Communications in Cognitive Radio Networks, energy harvesting in wireless network, D2D communication, and physical layer security issues in Wireless Networks, and UAV assisted communication. He has published more than 40 research papers in various journals and conferences. He is a reviewer of several IEEE, Springer, Wiley and Elsevier conferences and journals.

REFERENCES

1. M. Z. Chowdhury, M. Shahjalal, S. Ahmed, Y. M. Jang, 6G wireless communication systems: Applications, requirements, technologies, challenges, and research directions, IEEE Open Journal of the Communications Society 1 (2020) 957–975. doi:10.1109/OJCOMS.2020.3010270.
2. M. S. Mushtaq, A. Mellouk, B. Augustin, S. Fowler, QoE power-efficient multimedia delivery method for LTE-A, IEEE Systems Journal 10 (2) (2016) 749–760. doi:10.1109/JSYST.2015.2435994.
3. W. Jiang, B. Han, M. A. Habibi, H. D. Schotten, The road towards 6G: A comprehensive survey, IEEE Open Journal of the Communications Society 2 (2021) 334–366.
4. S. Talwar, N. Himayat, H. Nikopour, F. Xue, G. Wu, V. Ilderem, 6G: connectivity in the era of distributed intelligence, IEEE Communications Magazine 59 (11) (2021) 45–50. doi:10.1109/MCOM.011.2100162.
5. B. Li, P. Hou, H. Wu, F. Hou, Optimal edge server deployment and allocation strategy in 5G ultra-dense networking environments, Pervasive and Mobile Computing 72 (2021) 101312.
6. A. Kumar, S. K. Suman, L. Bhagyalakshmi, A. K. Sahu, Iot and cloud network based water quality monitoring system using IFTTT framework. In: Mahajan, V., Chowdhury, A., Padhy, N. P., Lezama, F. (eds) Sustainable Technology and Advanced Computing in Electrical Engineering. *Lecture Notes in Electrical Engineering.* Springer, Singapore. vol 939 (2022) 23–32. doi: 10.1007/978-981-19-4364-5_3.
7. M. B. H. Weiss, K. Werbach, D. C. Sicker, C. E. C. Bastidas, On applyingblockchains to spectrum management, IEEE Transactions on Cognitive Communications and Networking 5 (2) (2019) 193–205. doi:10.1109/TCCN.2019.2914052.
8. C.-H. Hsu, G. Manogaran, G. Srivastava, N. Chilamkurti, Guest editorial: 6G-enabled network in the box (NIB) for industrial applications and services, IEEE Transactions on Industrial Informatics 17 (10) (2021) 7141–7144. doi:10.1109/TII.2021.3067707.
9. V. Ziegler, S. Yrjl, How to make 6G a general purpose technology: Prerequisites and value creation paradigm shift. *Joint European Conference on Networks and Communications & 6G Summit (EuCNC/6G Summit)*, Porto, Portugal (2021) 586–591. doi:10.1109/EuCNC/6GSummit51104.2021.9482431.
10. M. Elsayed, M. Erol-Kantarci, AI-enabled future wireless networks: Challenges, opportunities, and open issues, IEEE Vehicular Technology Magazine 14 (3) (2019) 70–77.
11. L. Bonati, S. D'Oro, M. Polese, S. Basagni, and T. Melodia, "Intelligence and Learning in O-RAN for Data-driven NextG Cellular Networks," IEEE Communications Magazine, vol. 59, no. 10, pp. 21–27, October 2021.
12. O-RAN Working Group 1, "O-RAN Architecture Description 5.00," ORAN.WG1.O-RAN-Architecture-Description-v05.00 Technical Specification, July 2021.
13. O-RAN Working Group 2, "O-RAN AI/ML Workflow Description and Requirements 1.03," O-RAN.WG2.AIML-v01.03 Technical Specification, July 2021.
14. O-RAN Working Group 2, "O-RAN Non-RT RIC Architecture 1.0," O-RAN.WG2. Non-RT-RIC-ARCH-TS-v01.00 Technical Specification, July 2021.
15. O-RAN Working Group 3, "O-RAN Near-RT RAN Intelligent Controller Near-RT RIC Architecture 2.00," O-RAN.WG3.RICARCHv02.00, March 2021.
16. H. Lee, J. Cha, D. Kwon, M. Jeong, and I. Park, "Hosting AI/ML Workflows on O-RAN RIC Platform," in Proceedings of IEEE GLOBECOM Workshops, December 2020.
17. A. S. Abdalla, P. S. Upadhyaya, V. K. Shah, and V. Marojevic, "Toward Next Generation Open Radio Access Network–What O-RAN Can and Cannot Do!" arXiv preprint arXiv:2111.13754 [cs.NI], November 2021.

18. L. Liang, J. Kim, S. C. Jha, K. Sivanesan, G. Y. Li, Spectrum and power allocation for vehicular communications with delayed CSI feedback, IEEE Wireless Communications Letters 6 (4) (2017) 458–461

19. A. Garcia-Saavedra and X. Costa-Perez, "O-RAN: Disrupting the Virtualized RAN Ecosystem," IEEE Communications Standards Magazine, pp. 1–8, 2021.

20. B. Brik, K. Boutiba, and A. Ksentini, "Deep Learning for B5G Open Radio Access Network: Evolution, Survey, Case Studies, and Challenges," IEEE Open Journal of the Communications Society, pp. 1–1, January 2022.

21. G. Li, L. Yan, and B. Qu, "Multi-Objective Particle Swarm Optimization Based on Gaussian Sampling," IEEE Access, vol. 8, pp. 209717–209737, 2020, doi: 10.1109/ACCESS.2020.3038497.

4 Two-User Cooperative NOMA with a SWIPT-Enabled Relay

*Hritwika Sarkar, Shashibhushan
Sharma, and Sumit Kundu*

4.1 INTRODUCTION: NOMA FOR 5G AND BEYOND NETWORKS

Next-generation networks of 5G and beyond are evolving to provide connectivity and services to large number of devices. Spectral efficiency is one of the most important requirements to meet this demand. Non-orthogonal multiple access (NOMA) has drawn a lot of interest as a potentially useful technique for supporting a large number of devices with excellent spectrum efficiency [1, 2]. In power domain NOMA, users will be separated on the basis of power level at the receiver using successive interference cancellation (SIC). In the uplink NOMA communication, the highest power is assigned to the information signal of the user nearest to the base station, whereas the highest power is assigned to the information signal of the distant user from the base station in the case of downlink mode of NOMA communication [3, 4]. In the uplink NOMA, the information signal of the nearest user is decoded by the base station first and after SIC operation, the information signal of the distant user is decoded. In the downlink mode of communication, the nearest user first decodes the information signal of the distant user and then decodes its own information signal.

Currently, NOMA has attracted significant research attention. The performance of outage of a NOMA-based uplink land mobile satellite communications has been studied by the authors of [5]. The authors in [6] have analyzed massive machine-type communication, where the different users sending their data with different power levels utilize an uplink NOMA. Here the authors have shown the energy efficient massive connectivity. The authors in [7] have analyzed the device-to-device communication in an uplink NOMA-based communication. In this model, the far user takes the help of the nearest user to send the data. The ergodic rate was analyzed, which shows that the performance is better in comparison to the conventional orthogonal multiple access. The authors in [8] have analyzed the NOMA in satellite communication and shown that the insertion of NOMA in satellite communication is superior to other communication systems. To further enhance the performance of NOMA in terms of transmission reliability, extension of coverage, cooperative NOMA has been proposed where one user with good channel gain acts as a cooperative relay, or a separate dedicated node acts as a relay [2, 9, 10]. The authors in [11] have analyzed the NOMA-based downlink communication where

DOI: 10.1201/9781003407836-4

the satellite directly and with the help of earth-based relay communicates with the users. The outage probability has been analyzed and the performance is shown to be better than conventional schemes.

Simultaneous wireless information and power transfer (SWIPT) enables power-constrained relay nodes to harvest energy from the radio frequency (RF) signal and helps in achieving green communication [12]. There are two interesting protocols for harvesting energy such as time switching relaying (TSR) and power splitting relaying (PSR) [12]. Further, a hybrid of these two schemes can also be used. Energy-constrained relay in cooperative NOMA uses energy harvesting vide SWIPT. Outage of NOMA and SWIPT-based NOMA have been investigated in [13–15]. The authors in [16] have analyzed the outage probability and ergodic rate in downlink NOMA networks with energy-harvesting relay. Here, multiple relays harvest the energy from radio frequency of base station. A particular relay is partially selected among the multiple energy-harvesting relays to forward the information signals. In this network, the imperfect SIC operation and imperfect channel state information have been considered. In [17], the authors have analyzed the energy harvesting based on time splitting mechanisms. Here, all the transmitting nodes harvest energy from a power beacon. The authors have analyzed the outage probability, throughput, and energy efficiency as performance metrics. In [18], the authors have studied energy harvesting in NOMA-based downlink communication where a relay harvests energy from a power beacon in non-linear mode. Here, the authors used two types of antenna-selection schemes to improve the outage performance.

In this chapter, we investigate a model where two separate sources use NOMA to send signals to two different destinations with the aid of a middle energy limited decode and forward (DF) relay. The signals of the two senders help the relay to gather energy using two separate energy-harvesting schemes: (i) TSR and (ii) PSR.

The major contribution of our chapter shows the following:

 i. Probability of outage for two-user NOMA aided by a DF relay harvesting energy under (a) TSR scheme and (b) PSR scheme is presented analytically
 ii. Closed-form analytical equations for the likelihood of an outage in each of the two energy harvesting plans
iii. Showing how power allocation coefficients affect the outage
iv. Outlining how time switching and power splitting parameters affect system outages

MATLAB-based simulation results validated our analytical formulation.

4.2 NETWORK MODEL FOR TWO-USER NOMA

4.2.1 NETWORK MODEL

The two-hop NOMA system model with an energy-harvesting relay is shown in Figure 4.1. Two sources (S1 and S2) communicate with two destinations (D1 and D2): a near user and a remote user with the help of a relay. Here, half-duplex mode is used by each node, each of which has a single antenna, and the sources and the

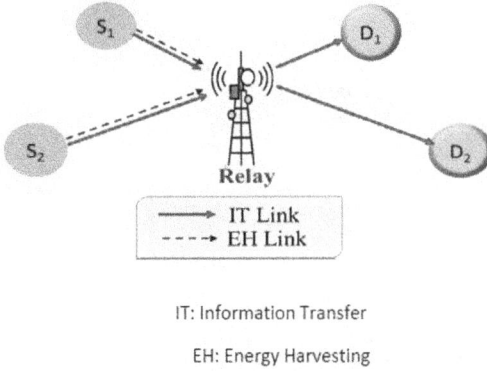

IT: Information Transfer

EH: Energy Harvesting

FIGURE 4.1 System model for two-user cooperative NOMA.

related destinations are not directly connected. Furthermore, quasi-static Rayleigh fading is assumed for all the channels. The signals transmitted by the sources help the relay to gather energy following (a) PSR and (b) TSR under two circumstances during the model's first hop, or uplink phase. In the downlink phase, which is the second hop of transmission, it uses that energy that was harvested to transmit the information from the source to the destinations. The relay initially utilizes SIC in the uplink to decode the signal of S1 and S2, then it employs superposition coding to create a NOMA signal and uses the collected energy to send it. Further, the nearest users D1 performs SIC in the downlink.

4.2.2 CHANNEL CHARACTERISTICS

All the channels considered in the current work are modelled as circularly symmetric complex Gaussian random variables with a mean of zero and a variance σ_x i.e. $CN(0, \sigma_x^2)$, where x denotes various channels. Additionally, we consider additive white Gaussian noise (AWGN) to be a complex Gaussian random variable with circular symmetry, having a mean of 0 and a variance of N_0, or $CN(0, N_0)$. h_{S_1R} and h_{S_2R} represent the coefficients associated with channels from sources to the relay. Next, h_{RD_1} and h_{RD_2} are used to denote the coefficients of channels from the relay to the users. These links have the channel gains, which are $g_{S_1R}, g_{S_2R}, g_{RD_1}$ and g_{R_2} respectively, where $g_i = |h_i|^2$. These channel gains are considered to be exponential random variables with independent and identical distributions (i.i.d) and a mean channel gain of $2\sigma_x^2 (= \lambda_x)$. The coefficients of exponential channel gain are $\lambda_{S_1R}, \lambda_{S_2R}, \lambda_{RD_1}$ and λ_{RD_2} respectively [8].

4.2.3 ENERGY HARVESTING AT THE RELAY

The period of harvesting energy and forwarding the signal is shown in Figure 4.2 for both the PSR and TSR scheme.

The TSR scheme divides the total time (T) used in communication into three parts. During the initial time slot, τT, the transmitted signals of two sources, helps

T

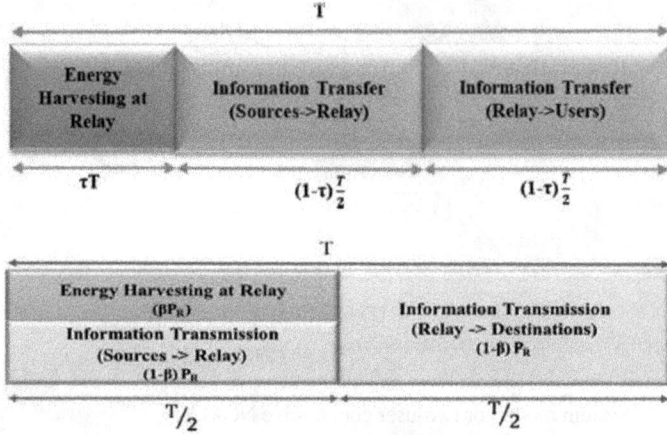

FIGURE 4.2   (a) Time frame for TSR scheme. (b) Time frame for PSR scheme.

the relay to harvest energy, where τ is the time switching factor $\in (0,1)$ [5]. The obtained energy is provided as

$$E_h = \tau T \left(P_{S_1}\, g_{S_1R} + P_{S_2}\, g_{S_2R} \right). \tag{4.1}$$

The information from the sources is received at the relay in the second time slot of $(1-\tau)\frac{T}{2}$. The relay transmits the signal with harvested energy at a power of $P_R = \frac{E_h}{(1-\tau)T/2}$ in the last time slot of $(1-\tau)\frac{T}{2}$.

According to PSR scheme, β (power splitting factor) is the fraction of source power utilized for harvesting energy $\beta \in (0,1)$ [5]. In half of the time slot T/2, relay harvests energy from the RF signals of two sources with a fraction of β of total received power, while the information is received from sources to relay with a fraction power of $(1-\beta)$,

$$E_h = \beta \left(\frac{T}{2} \right) \left(P_{S_1}\, g_{S_1R} + P_{S_2}\, g_{S_2R} \right), \tag{4.2}$$

where P_{S_1}, P_{S_2} are the transmit power of S_1 and S_2 respectively. Here g_{S_1R} and g_{S_2R} are the channel gains of source to relay. The signal is decoded and forwarded by relay in the third time slot, $\frac{T}{2}$. Relay utilizes the harvested energy and the corresponding power for transmitting the signal is $P_R = \frac{E_h}{T/2}$.

4.3   OUTAGE ANALYSIS WITH TSR AND PSR SCHEME AT THE RELAY

Two phases of the communication are accomplished. The signals from the sources are sent to the relay during the first phase. The signal sent over the air is written as

$$x = \sqrt{P_{S_1}}\, x_{S_1} + \sqrt{P_{S_2}}\, x_{S_2}, \tag{4.3}$$

where P_{S_1} and P_{S_2} are the allocated power for S1 and S2 for the symbols x_{S_1} and x_{S_2}.

The received signal at relay for TSR scheme

$$y_R = \sqrt{P_{S_1}}\, h_{S_1 R} x_{S_1} + \sqrt{P_{S_2}}\, h_{S_2 R} x_{S_2} + N_0 \ \text{ where } P_{S_1} > P_{S_2}. \tag{4.4}$$

The received signal at relay for PSR scheme

$$y_R = \sqrt{(1-\beta)P_{S_1}}\, h_{S_1 R} x_{S_1} + \sqrt{(1-\beta)P_{S_2}}\, h_{S_2 R} X_{S_2} + N_0 \ \text{ where } P_{S_1} > P_{S_2}. \tag{4.5}$$

Relay uses gathered energy (by harvesting) to broadcast the successfully decoded messages x_{S_1} and x_{S_2} to users 1 and 2 respectively. The received signals at user1 and user2 for both TS and PS methods are

$$y_{RD_1} = \sqrt{\alpha_1 P_R}\, h_{RD_1} x_{S_1} + \sqrt{\alpha_2 P_R} h_{R_1} x_{S_2} + N_0 \ \text{ and} \tag{4.6}$$

$$y_{RD_2} = \sqrt{\alpha_1 P_R}\, h_{RD_2} x_{S_1} + \sqrt{\alpha_2 P_R} h_{R_2} x_{S_2} + N_0, \tag{4.7}$$

where N_0 denotes AWGN. Here, α_1 is the user1 power allocation coefficient, while α_2 is the user2 power allocation coefficient. $\alpha_1 + \alpha_2 = 1$, $\alpha_2 > \alpha_1$. In the first hop of the TS technique, the signal to interference-plus-noise ratio (SINR) of information signals [8]

$$\gamma_{S_1 R}^1 = \frac{P_{S_1} g_{S_1 R}}{P_{S_2} g_{S_2 R} + N_0}, \ \gamma_{S_2 R}^2 = \frac{P_{S_2} g_{S_2 R}}{N_0}. \tag{4.8}$$

In the first hop of the PS method, the information signal's SINR is expressed as

$$\gamma_{S_1 R}^1 = \frac{(1-\beta)P_{S_1} g_{S_1 R}}{(1-\beta)P_{S_2} g_{S_2 R} + N_0}, \ \gamma_{S_2 R}^2 = \frac{(1-\beta)P_{S_2} g_{S_2 R}}{N_0}. \tag{4.9}$$

As D_1 is nearer, SIC occurs at D_1 first; D_2 is decoded first considering D_1 as interference. In the downlink, the SINR of D_1 after decoding and subtracting D_2 in TS and PS is

$$\gamma_{RD_1}^1 = \frac{\alpha_1 P_R g_{RD_1}}{N_0}. \tag{4.10}$$

The SINR of user2 at D_2 is

$$\gamma_{RD_2}^2 = \frac{\alpha_2 P_R g_{RD_2}}{\alpha_1 P_R g_{RD_2} + N_0}. \tag{4.11}$$

P_R is the transmit power of the relay based on harvested energy corresponding to TS and PS schemes.

The overall SNR of user1 and user2 in case of DF relay is expressed by

$$\gamma_{S_1 D_1}^1 = min\left(\gamma_{S_1 R}^1, \gamma_{RD_1}^1\right) \gamma_{S_2 D_2}^2 = min\left(\gamma_{S_2 R}^2, \gamma_{RD_2}^2\right). \tag{4.12}$$

Decoded signal capacities at D1 for the signal from S1 and at D2 for the signal from S2 in TS are [8]

$$C_{S_1 D_1}^1 = \frac{1-\tau}{2} log_2\left(1+\gamma_{S_1 D_1}^1\right) \text{ and, } C_{S_2 D_2}^2 = \frac{1-\tau}{2} log_2\left(1+\gamma_{S_2 D_2}^2\right), \text{ respectively.} \quad (4.13)$$

Corresponding capacities in PS relaying method are

$$C_{S_1 D_1}^1 = \frac{1}{2} log_2\left(1+\gamma_{S_1 D_1}^1\right) C_{S_2 D_2}^2 = \frac{1}{2} log_2\left(1+\gamma_{S_2 D_2}^2\right). \quad (4.14)$$

P_{OUT} stands for the probability of an end-to-end system failure, and r_{th_1} and r_{th_2} are the target rates for users 1 and 2, respectively:

$$P_{OUT} = P\left\{\left(C_{S_1 D_1}^1 < r_{th_1}\right) \cup \left(C_{S_2 D_2}^2 < r_{th_2}\right)\right\}. \quad (4.15)$$

For TSR scheme the outage is given as [7]

$$
\begin{aligned}
P_{OUT} &= 1 - P\left(\gamma_{S_1 R}^1 > \theta_1\right) P\left(\gamma_{RD_1}^1 > \theta_1\right) P\left(\gamma_{S_2 R}^2 > \theta_2\right) P\left(\gamma_{RD_2}^2 > \theta_2\right) \\
&= 1 - \left(\frac{e^{-\theta_1 K_1}}{1+K_2 \theta_1}\right)\left(e^{-\frac{\theta_1 K_3}{\alpha_1}}\right)\left(e^{-\theta_2 K_4}\right)\left(e^{-\frac{K_5 \theta_2}{(\alpha_2 - \alpha_1 \theta_2)}}\right),
\end{aligned} \quad (4.16)
$$

where $\theta_1 = \left(2^{\frac{2r_{th_1}}{1-\tau}}\right) - 1$ $\theta_2 = \left(2^{\frac{2r_{th_2}}{1-\tau}}\right) - 1$ and $K_1 = \frac{N_0}{P_{S1}\lambda_{S1R}}$, $K_2 = \frac{P_{S2}\lambda_{S2R}}{P_{S1}\lambda_{S1R}}$, $K_3 = \frac{N_0}{P_R \lambda_{RD1}}$, $K_4 = K_1 / K_2$ and $K_5 = \frac{N_0}{P_R \lambda_{RD2}}$.

For PSR, the overall outage is

$$
\begin{aligned}
P_{OUT} &= 1 - P\left(\gamma_{S_1 R}^1 > \theta_1\right) P\left(\gamma_{RD_1}^1 > \theta_1\right) P\left(\gamma_{S_2 R}^2 > \theta_2\right) P\left(\gamma_{RD_2}^2 > \theta_2\right) \\
&= 1 - \left(\frac{e^{-\theta_1 K_1}}{1+K_2 \theta_1}\right)\left(e^{-\theta_1 K_3}\right)\left(e^{-\theta_2 K_4}\right)\left(e^{-\left(\frac{1}{K_5 - \frac{K_6}{\theta_2}}\right)}\right),
\end{aligned} \quad (4.17)
$$

where $K_1 = \frac{N_0}{(1-\beta)P_{S1}\lambda_{S1R}}$, $K_2 = \frac{P_{S2}\lambda_{S2R}}{P_{S1}\lambda_{S1R}}$, $K_3 = \frac{N_0}{\alpha_1 P_R \lambda_{RD1}}$, $K_4 = K_1 / K_2$ and $K_5 = \frac{\alpha_1 P_R}{N_0 \lambda_{RD2}}$

$$\theta_1 = 2^{2r_{th_1}} - 1 \text{ and } \theta_2 = 2^{2r_{th_2}} - 1.$$

4.4   RESULTS AND DISCUSSION

4.4.1   RESULTS WITH TSR SCHEME OF RELAYING

We now present the system outage probability (SOP) for both the harvesting schemes for various parameters following [15]. Transmit power of source S_1 (P_{s1}) and source S_2 (P_{s2}): 5 dBW, 10 dBW, 15 dBW; coefficient of time sitching (τ): 0-1; efficiency of energy harvesting (η): 0.8; NOMA coefficient for fractional relay power (α_1): 0-1; noise power of AWGN (N_o) 0.1 dB W; threshold rates for outage (r_{th1}, r_{th2}): 0.02 b/s/Hz,

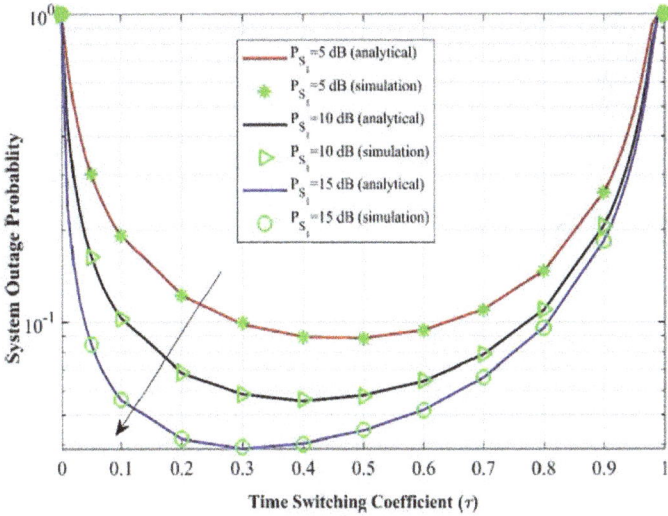

FIGURE 4.3   Outage probability versus time switching coefficient for energy harvesting (τ) with different values of signal power of user1.

(Reprinted from [15] with permission.)

0.01 b/s/Hz, for Figures 4.3, 4.4, 4.6, and 4.7, mean values of channel gain for near user (Ωi) and far user (Ωj) are 1, 0 and 0.25 respectively. We first present the performance of a TSR-based harvesting scheme [15].

Figure 4.3 displays the SOP for the EH relay-aided NOMA network for several values of user1's SNR with variation in EH time (τ) in the TSR protocol. Based on the simulation and analytical result, we find that the optimal value of τ=0.4, 0.35, and 0.3 when PS1 = 5 dB, 10 dB, and 15 dB respectively. As energy is harvested more, it is observed that the likelihood of an outage decreases. Relay transmit power rises, and the outage value decreases. As τ rises, the capacity decreases due to a gradual decrease in transmission time of $(1-\tau)\frac{T}{2}$. As a result, we arrive at an ideal of τ where the outage chance is lowest.

Outage performance is depicted in Figure 4.4 in relation to the fraction of relay power allotted to user1 at the time when signals are delivered from the relay to destinations. According to the downlink NOMA procedures, user1, who is closer to the relay, receives low power allocation, while user2, who is farther away, receives higher power allocation. So, D1 gets lower power while D2 gets more power for decoding. It may be noted that when relay to D1 and relay to D2 channel capacity become less than the desired threshold causing both the signals to go to link outage, a system outage occurs. System outage is getting improved with the increasing value of $\alpha1$ as channel strength is getting better for D1 and D2. However, after $\alpha1 > 0.5$, their overall system performance becomes worse. As the near user is getting higher power, so it does not go into outage. However, as $\alpha2$ becomes less valuable due to $\alpha2 < 0.5$, user2 goes into outage due to insufficient power for decoding. Therefore, the likelihood of an outage is very high for both the very high and very low values of $\alpha1$.

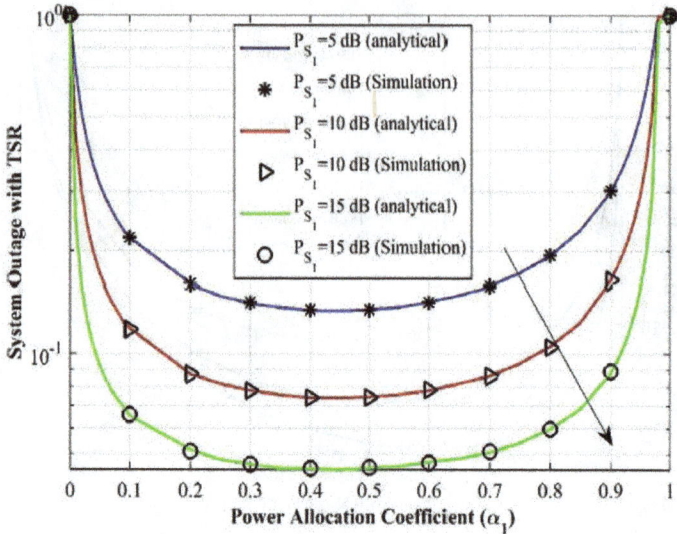

FIGURE 4.4 Outage probability versus power allocation coefficient from relay (α1) with different values of ρ_{S1}.

(Reprinted from [15] with permission.)

From this simulation, we are able to determine the optimal relay power fraction at α1=0.4 when PS1 was 5, 10, or 15 dBW. Further, as α1 approaches to 1, outage is seen to approach 1.

The outage probability with respect to SNR of S2 for various values of data rates is shown in Figure 4.5. We see that the system initially had a decreasing outage probability when the transmit SNR of S2 is increased gradually. Additionally, we obtain an SNR for S2 that is optimal, at which point outage is minimum. Further, as SNR for S2 increases, outage increases until it reaches 1. In order to have better performance with respect to outage, S1 transmits message signals at a high power level, while S2 transmits at a reduced power level. Performance during outages declines as threshold rate values rise. Relay can successfully decode for transmission when the source power is strong and the threshold rate is low. As a result, outage performance increases as the threshold is lowered. The outcomes of the simulation are observed to very closely agree with the outcomes of the analysis.

4.4.2   RESULTS WITH PSR SCHEME OF RELAYING

Figure 4.6 displays the energy harvesting network's outage probability in relation to the power splitting coefficient (β) for various user1 SNR values. We got an optimal PSR parameter of $\beta = 0.6$ when $P_{S1} = 5$ dB, $\beta = 0.5$ when $P_{S1} = 10$ dB, and $\beta = 0.4$ when $P_{S1} = 15$ dB from the simulation. The harvesting energy increases with increase in β. It shows improvement in outage performance with increase in β as the capacity increases due to increase in harvesting energy. Consequently, transmit

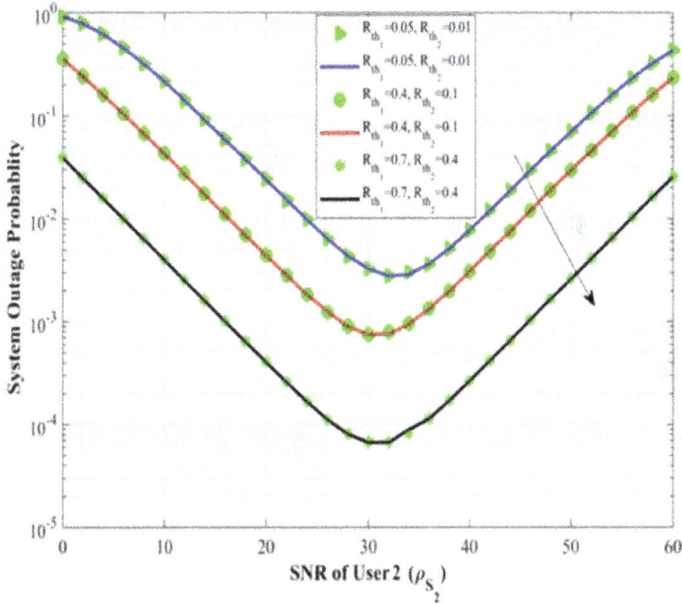

FIGURE 4.5 SOP at varied threshold rates with the influence of source2's SNR. (Reprinted from [15] with permission.)

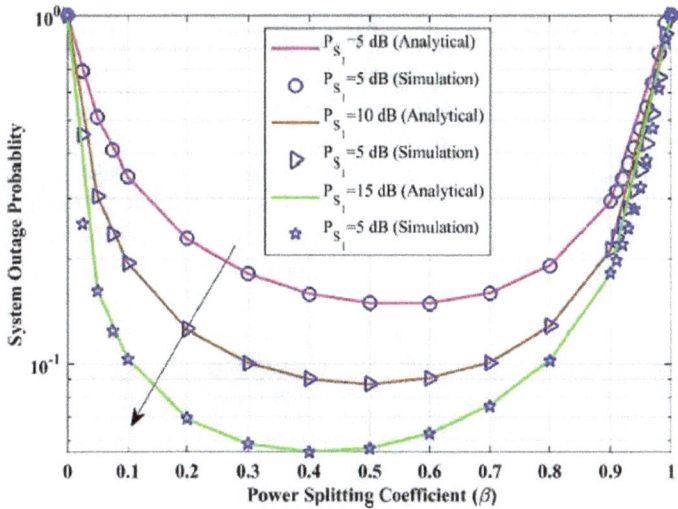

FIGURE 4.6 Outage probability versus β with different values of power of source1.

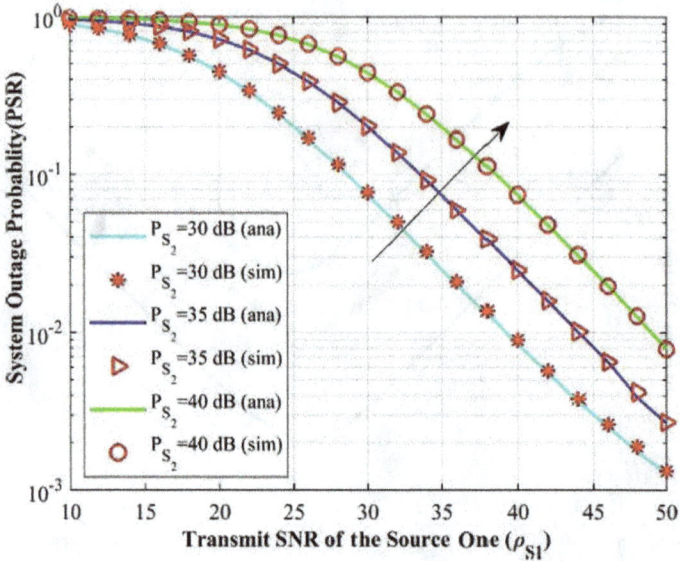

FIGURE 4.7 Outage probability versus SNR of source1 (ρ_{S1}) with different values of power of source2 (ρ_{S2}).

power increases which reduces outage. After the optimum value of β, outage again increases. As there are further increase in β, though it increases the harvested energy, it reduces the power $(1-\beta)$ with which information from source to relay is received, which therefore reduces the SNR received at the relay. Also, it can be observed that increasing the source power which leads to higher SINR at destination tends to improve outage performance.

The nature of SOP curve is convex, due to the fact that with increase in value of β, the harvested energy is getting better; however, after an optimal point of β, performance for transmission is getting worse, as power is not sufficient for decoding, which in turn causes outage probability to tend to 1.

Figure 4.7 shows SOP with respect to transmitting SNR of user1 (ρS1) through simulation results which are verified by analytical expressions. In the uplink stage, SNR of first source is more than second source. The relay first decodes the signal of first source cancelling the second one by considering it as an interference signal. As ρS1 increases, outage probability decreases. In downlink stage, relay transmits the signal to receivers by power allocating α1 and α2. As α1 < α2, user cancels out user2's interference using SIC techniques to retrieve user1's signal. Because signal 2 interferes with signal 1, the likelihood of an outage decreases with increasing signal power of user1, while the likelihood of an outage increases with increasing signal power of user2.

For various levels of data rates, Figure 4.8 displays outage probability relative to SNR of source2. When the SNR transmitted from the source2 increases, the network gradually experiences low outage probabilities, we reach an optimal value. Further, as the SNR increases, the outage probability increases to 1. It is seen that

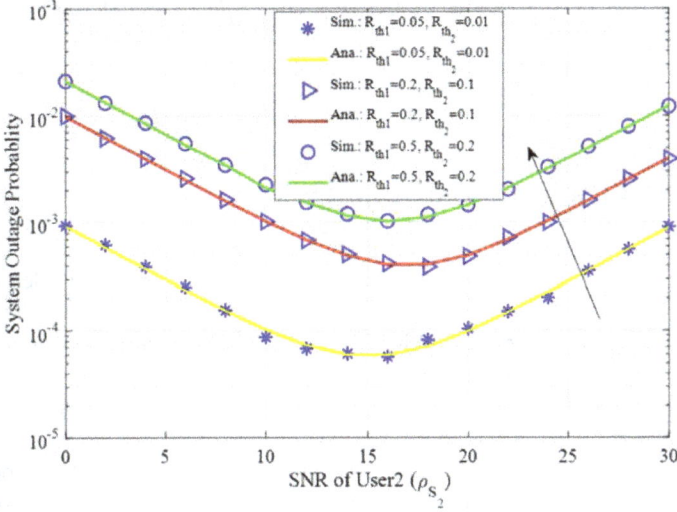

FIGURE 4.8 SOP versus SNR of source1 with different values of power of source2.

when source1 transmits message with higher power compared to user2, better outage performance is obtained. As the detection threshold value rises, outage performance falls. With lower value of detection threshold, probability of successful decoding of sources increases, causing improvement in outage performance. When the source power is high while the detection threshold is low, it allows the relay to capture more energy for transmission and chance of correct decoding increases. So outage performance becomes better; however, by increasing threshold, successful decoding of the signal gets reduced. So outage probability becomes higher.

4.5 CONCLUDING REMARKS

Outage performance of a double hop communication utilizing an intermediate DF relay is analyzed with two protocols for harvesting energy: TS- and PS-relaying protocols. The relay used in the system model is half duplex in nature. We observe optimal energy harvesting coefficient at which SOP attains minimum value. We can observe that when energy harvesting parameter in PS (i.e., power splitting ratio is less than 0.5), performance is better because the relay becomes able to harvest enough energy from both the sources for further transmission. But when it gets very high, though the relay can gather more energy, the information is not received with high power at the relay. We also observe that fraction of relay power for decoding the signals to desired destinations has some optimal value, which is between 0.4–0.6. Similarly in the case of TSR, an optimal TS parameter is observed, which minimizes outage. As the near user is having better channel strength, so it can be decoded with comparatively lesser power than the far user, which has low channel strength. So, power allocation factor should be always more for D2 than D1 for better outage. We can also observe the outage performance with the variation of SNR of source1.

At the low value of SNR, outage of both are high as some power is used for energy harvesting, and less power is involved for transmission. Increase in target data rate degrades outage performance. The study is significant in designing a double hop communication based on NOMA.

AUTHORS' BIOGRAPHICAL DETAILS

Hritwika Sarkar received her B.Tech in Electronics & Communication Engg from Haldia Institute of Technology (under MAKAUT) in 2017 and M.Tech in Telecommunications Engineering from the National Institute of Technology, Durgapur in 2022. Her Masters thesis is on "Performance of Dual-hop CR-NOMA Network with Energy Harvesting Relays." Her areas of research interest include cooperative relaying, NOMA for 5G, cognitive radio networks, and energy harvesting. She is currently associated with Cadence Design Systems since July 2022 in the role of Product Validation Engineer and takes interest in VLSI design. She is responsible for qualifying Quantus (cadence sign-off extraction tool).

Shashi Bhushan Sharma received his B.E. degree in Electronics and Communication Engineering from the University of Rajiv Gandhi Proudyogiki Vishwavidyalaya, India in 2011 and his M.Tech. degree in Telecommunication Engineering from the National Institute of Technology, Durgapur, India in 2015. He received his Ph.D. in Wireless Communication from the Department of Electronics and Communication Engineering, National Institute of Technology Durgapur, India in 2021. He has worked as a Post-Doctoral researcher in IIT Kharagpur in Department of Computer Science and Engineering in Sessions 2021–2022 and 2022–2023. Currently, he is working as an assistant professor in the Department of Electronics and Communication of Allahabad University. His research areas are physical layer security, cognitive radio network, two-way communication with half and full duplex relay, and energy harvesting.

Sumit Kundu received his Bachelor of Engineering (Hons) in Electronics & Communication Engineering from the National Institute of Technology, Durgapur (Erstwhile Regional Engg College, Durgapur, University of Burdwan) with a university gold medal in 1991, Master of Technology (M.Tech) in Telecommunication Systems Engineering from IIT Kharagpur in 1993–94, and Ph.D. (Wireless Communication) from IIT Kharagpur in 2004. He has been a faculty in the department of ECE, NIT Durgapur since

1995, where he is currently a full-time professor (Higher Administrative Grade). He also served as an assistant professor in the G.S.Sanyal School of Telecommunications, IIT Kharagpur for a year in 2007. He has supervised 11 Ph.D.s so far in the domain of wireless communication and networking and continuing supervision for several more students, He has published more than 200 research papers in international journals and national and international conferences. He has also contributed several book chapters. His current areas of research include cooperative communication, NOMA for 5G, intelligent reflecting surfaces (IRS), cognitive radio networks, energy harvesting in wireless networks, physical layer security, and wireless sensor networks. He is a senior member of IEEE. He is a reviewer of several papers for IEEE, Elsevier Journals, and national and international IEEE Conferences.

REFERENCES

1. Y. Liu, Z. Qin, M. Elkashlan, Z. Ding, A. Nallanathan and L. Hanzo, "Non-orthogonal Multiple Access for 5G and Beyond," in Proceedings of the IEEE, vol. 105, no. 12, pp. 2347–2381, Dec. 2017.
2. Abhinav Singh Parihar, Pragya Swami, Vimal Bhatia, and Zhiguo Ding, "Performance Analysis of SWIPT Enabled Cooperative-NOMA in Heterogeneous Networks Using Carrier Sensing", in IEEE Transactions on Vehicular Technology, vol. 70, no. 10, pp. 10646–10656, October 2021.
3. Mohammad Salehi, Hina Tabassum and Ekram Hossain, "Meta Distribution of SIR in Large Scale Uplink and Downlink NOMA Networks", in IEEE Transaction on Communications, vol. 67, no. 4, pp. 3009–3025, April 2019, doi: 10.1109/TCOMM.2018.2889484.
4. Zhiguo Ding, Yuanwei Liu, Jinho Choi, Qi Sun, Maged Elkashlan, I Chih-Lin and H Vincent Poor, "Application of Non Orthogonal Multiple Access in LTE and 5G Networks", in IEEE Communications Magazine, vol. 55, no. 2, pp. 185–191, February 2017, doi: 10.1109/MCOM.2017.1500657CM.
5. S. A. Tegos, P. D. Diamantoulakis, J. Xia, L. Fan and G. K. Karagiannidis, "Outage Performance of Uplink NOMA in Land Mobile Satellite Communications," in IEEE Wireless Communications Letters, vol. 9, no. 10, pp. 1710–1714, Oct. 2020, doi: 10.1109/LWC.2020.3001916.
6. M. B. Shahab, S. J. Johnson, M. Shirvanimoghaddam, M. Chafii, E. Basar and M. Dohler, "Index Modulation Aided Uplink NOMA for Massive Machine Type Communications," in IEEE Wireless Communications Letters, vol. 9, no. 12, pp. 2159–2162, Dec. 2020, doi: 10.1109/LWC.2020.3015920.
7. M. F. Kader, S. M. R. Islam and O. A. Dobre, "Simultaneous Cellular and D2D Communications Exploiting Cooperative Uplink NOMA," in IEEE Communications Letters, vol. 25, no. 6, pp. 1848–1852, June 2021, doi: 10.1109/LCOMM.2021.3062111.
8. X. Yan, H. Xiao, C. -X. Wang and K. An, "Outage Performance of NOMA-Based Hybrid Satellite-Terrestrial Relay Networks," in IEEE Wireless Communications Letters, vol. 7, no. 4, pp. 538–541, August 2018, doi: 10.1109/LWC.2018.2793916.
9. Zhiyuan Yu, Chao Zhai, Ju Liu and Hongji Xu," Cooperative Relaying Based Non-Orthogonal Multiple Access(NOMA) with Relay Selection," in IEEE Transaction on Vehicular Technology, vol. 67, no. 12, December 2018, pp. 11606–11618.
10. Yuehao Yu, Zheng Yang, Yi Wu, Jamal Ahmed Hussein, Wen-Kang Jia, and Zhicheng Dong, "Outage Performance of NOMA in Cooperative Cognitive Radio Networks With SWIPT," IEEE Access, vol. 7, 2019 pp. 117308–117317.

11. X. Li, Y. Chen, P. Xue, G. Lv and M. Shu, "Outage Performance for Satellite-Assisted Cooperative NOMA Systems With Coordinated Direct and Relay Transmission," in IEEE Communications Letters, vol. 24, no. 10, pp. 2285–2289, October 2020, doi: 10.1109/LCOMM.2020.3004413.
12. A. A. Nasir, X. Zhou, S. Durrani and R. A. Kennedy," Relaying Protocols for Wireless Energy Harvesting and Information Processing," IEEE Transactions on Wireless Communications, vol. 12, no. 7, July 2013, pp. 3622–3636.
13. Yingting Liu, Yinghui Ye, Haiyang Ding, Feng Gao, and Hongwu Yang "Outage Performance Analysis for SWIPT-Based Incremental Cooperative NOMA Networks With Non-Linear Harvester," in IEEE Communications Letters, vol. 24, no. 2, February 2020, pp. 287–291.
14. Yuanwei Liu, Zhiguo Ding, Maged Elkashlan and H. Vincent Poor," Cooperative Non-orthogonal Multiple Access With Simultaneous Wireless Information and Power Transfer," in IEEE Journal on Selected Areas inCommunications, vol. 34, no. 4, pp. 938–953, April 2016.
15. Hritwika Sarkar, Shashibhushan Sharma and Sumit Kundu, "Outage Performance of Two Users Cooperative NOMA with Energy Harvesting Relay," in IEEE INDICON, November 2022.
16. S. Bisen, P. Shaik and V. Bhatia, "On Performance of Energy Harvested Cooperative NOMA Under Imperfect CSI and Imperfect SIC," in IEEE Transactions on Vehicular Technology, vol. 70, no. 9, pp. 8993–9005, September 2021, doi: 10.1109/TVT.2021.3099067.
17. T.-H. Vu and S. Kim, "Performance Evaluation of Power-Beacon-Assisted Wireless-Powered NOMA IoT-Based Systems," in IEEE Internet of Things Journal, vol. 8, no. 14, pp. 11655–11665, 15 July, 2021, doi: 10.1109/JIOT.2021.3058680.
18. L. Zhang, J. Zhang, N. Hu, X. Li and G. Pan, "Outage Performance for NOMA-Based FSO-RF Systems with Transmit Antenna Selection and Nonlinear Energy Harvesting," in IEEE Internet of Things Journal, vol. 10, no. 7, pp. 6491–6506, 1 April, 2023, doi: 10.1109/JIOT.2022.3227043.

5 Non-Orthogonal Multiple Access (NOMA) for 5G Networks

Kirti Prakash, Mohd Javed Khan, and Ram Chandra Singh Chauhan

5.1 INTRODUCTION

From first-generation (1G) to fifth-generation (5G) wireless networks, a multiple access strategy is a key technology for differentiating distinct wireless systems. In 1G, 2G, 3G, and 4G, respectively, frequency division multiple access (FDMA), time division multiple access (TDMA), code division multiple access (CDMA), as well as orthogonal frequency division multiple access (OFDMA) were employed. These methods use orthogonal multiple access (OMA). To avoid or lessen inter-user interference, distinct users are allocated in an orthogonal manner within these multiple access structures, depending on the time, code, or frequency domain. In OMA, the deployment of suitably cost-effective receivers with minimal complexity makes it simple to isolate the information-carrying signals of the consumers. Nevertheless, due to orthogonal resources, OMA can handle a small number of users. The idea of non-orthogonal multiple access (NOMA) is put forth in order to support additional users in 5G [1].

On one hand, the use of innovative wireless technologies is also increasing the amount of mobile traffic by several thousand times compared to the remaining networks, while the upcoming 5G as well as next-generation systems are anticipated to provide trustworthy, flawless, and energy-efficient support with worldwide connectivity [2].

A potential multiple-access method for 5G and afterwards wireless communication networks is NOMA. Its goals are to effectively satisfy customer experience requirements and offer a theoretical framework for the future implementation of cells with reduced coverage. By accommodating many users in a single orthogonal resource piece, NOMA techniques achieve their goal by utilizing non-orthogonality and overloading. Although these ideas are not novel in 3G or 4G interactions, like CDMA, NOMA utilizes a non-orthogonal strategy at the center of its architectural structure and seeks a larger OC [3].

Intelligent transportation system (ITS) uses have exploded because of the rapid spread of self-driving vehicles equipped with built-in communication and sensor capabilities. The ITS solutions can provide enhanced data services, entertainment services, vehicle platooning, as well as on-road security and control of traffic.

DOI: 10.1201/9781003407836-5

Vehicle-to-vehicle (V2V) interactions are an essential component of 5G and beyond communication systems as a result of the enormous potential of NOMA techniques [4].

It has been greatly recognized how NOMA technologies have facilitated the 5G requirements for vast connection, ultra-high capability, rapid data transmission, ultra-low delay, and strong interference management. On an individual resource element (RE), NOMA multiplexes various users in the power or code areas [5]. It is seen as a promising solution for beyond 5G (B5G) wireless transmission because of the spectrum efficiency (SE) improvement over the prevailing technique [6].

In recent times, NOMA has drawn a lot of attention due to its promising use in 5G wireless networks. It is used to enhance the SE of the system. Users of the NOMA system may use consecutive interference cancelers to separate their own data from that of other individuals when the channel circumstances are favorable. However, this has the unwanted consequence of reducing the SE of the system by forcing NOMA to assign higher power to a customer with poor channel circumstances in order to correctly decode the overlaid signal [7].

5.2   NOMA SCHEMES

NOMA is a multi-access technology that enables users to utilize resources simultaneously in a non-orthogonal manner on a subcarrier by simultaneously broadcasting data on the exact same frequency with different levels of power or by code assignment [8, 9]. The reduction in latency seen during concurrent transmission is significantly impacted by NOMA. This indicates that the asset is best utilized by consuming all available bandwidth, which enhances the SE. Recent years have seen a significant increase in interest in NOMA schemes for 5G mobile phone networks. The primary justification for NOMA's adoption in 5G as well as 6G is its capacity to accommodate a large number of users while using same-time as well as frequency sources. In order to multiplex consumers in the power domain, NOMA uses overlay encoding in the transmitter as well as successive interference cancellation (SIC) at the receiver (Figure 5.1).

NOMA can be categorized into main types.

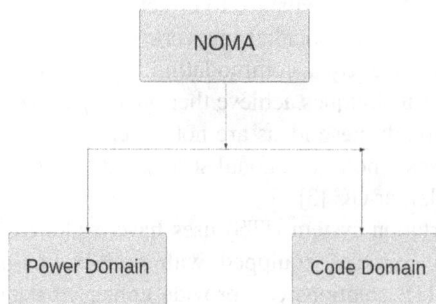

FIGURE 5.1   NOMA techniques.

5.2.1 Power-domain NOMA

A potential multiple-access (MA) technique for networks supporting 5G is power-domain NOMA [10]. For the 3rd Generation Partnership Project (3GPP) long-term evolution advanced (3GPP-LTE-A) systems, a downlink variant of NOMA termed multiuser superposition transmission (MUST) has been proposed [11]. It has been demonstrated that NOMA can increase system performance along with user satisfaction. More recently, 3GPP LTE Release 14 approved a new work article describing downstream for LTE [12], with the goal of identifying the appropriate techniques for LTE to implement the downstream intra-cell MUST. By identifying them with various power levels, power-domain NOMA supports numerous users inside the same time/frequency/code source block. Partial non-orthogonal multiple access (P-NOMA) often contains just one observation, in contrast to multiuser recognition in CDMA/multiple-input multiple-output (MIMO) frameworks, which typically have many observations at the point of reception.

5.2.2 Code-Domain NOMA

Multiple users can superimpose the signals side by side in the P-NOMA. The improvement in interference cancellation techniques is advantageous, while it is still challenging [13]. Thus, a NOMA code domain is the best option. By using unique sequences with limited, minimal-density and poor cross-correlation characteristics for every client, a NOMA multiplexes users.

5.2.2.1 Interleave Division Multiple Access (IDMA)

In the code-domain NOMA, the IDMA is regarded as a viable candidate. This particular instance of CDMA direct patterns may be taken into consideration. Rather than using CDMA to distinguish users, IDMA uses particular sequences. User-specific interleavers are another special feature employed in IDMA to separate users [14].

5.2.2.2 Multiuser Shared Access (MUSA)

MUSA [15], coupled with the overload, without grant access, is another important component. This technique fails to organize the resource base because it gives each user complete freedom to select from the extensive cardinality of their overflow sequence. By implementing grant-free accessibility, MUSA can lessen the overhead signaling as well as delay transmission brought on the traditional grant-based broadcasts. MUSA suggests the uplink communication since it lowers the energy usage of its devices. To support several free users and lessen the effects of a user collision, MUSA spreading patterns should be short and have low cross-correlation.

5.2.2.3 Sparse Code Multiple Access (SCMA)

A well-known cooperative NOMA (C-NOMA) solution being studied for 5G is called SCMA [16], in which a constrained number of users let one another interfere. Each user is given a particular, limited, multidimensional, and intricate codebook. To distribute the modulated patterns from individuals over the allotted resources in SCMA, codebooks are utilized to simultaneously recognize many users.

5.2.2.4 Low-density Spreading CDMA (LDS-CDMA)

LDS-CDMA, which is based on the original CDMA idea, uses LDS instead of traditional spreading patterns to reduce the degree of interference that typical CDMA schemes impose on each chip. The fundamentals of LDS-CDMA have been covered in [17, 18]. The repeated multi-user detection (MUD) built on the message-passing algorithm (MPA), which imposes a lower complexity compared to that of optimized MAP detector, was also explored [19]. The effectiveness of LDS-CDMA communication across memory-less Gaussian networks utilizing binary phase-shift keying (BPSK) modulation was specifically examined in [17].

5.3 NOMA WORKFLOW

Superposition coding (SC) is a technique used by the NOMA system to combine the various signals from all the transmitting individuals into a single signal. Signals are aggregated at the base unit in uplink NOMA. The users receive overlaid signals provided by the base station (BS) for downlink NOMA. SC at the point of transmission as well as SIC for the reception point enables each user to benefit from the same bandwidth. At the sender location, numerous data signals come together into only one waveform, whereas the SIC analyzes all of them separately until it discovers the signal that is wanted at the destination [20].

5.3.1 DOWNLINK NOMA TRANSMISSION

The BS transmits the sum of the message, which is a combination of the intended signals of various users with varied allotted power values, to all mobile subscribers from the transmitter end of the downstream NOMA system. It is believed that the SIC process will be carried out sequentially at each user's receiver till the user's information is retrieved. Users' power factors are distributed inversely proportionally to their channel conditions. A larger transmission power is allotted to the user who has a poor channel condition than to the user who has a better channel condition. As a result, the user with the most transmission power recovers its signal right away without going through any SIC processes because it perceives the signals of other devices as noise. But other users must carry out SIC operations [21]. In SIC, every user's receiver first finds signals which are stronger than the signal they want to receive. After that, the received signal is further removed from those signals until the associated user's unique signal is found (Figure 5.2).

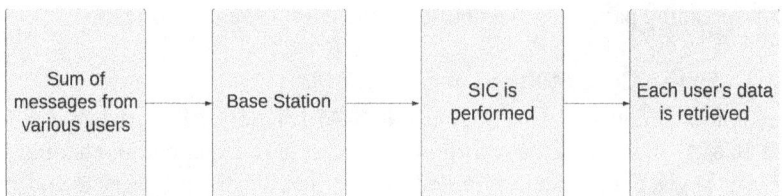

FIGURE 5.2 Downlink NOMA transmission.

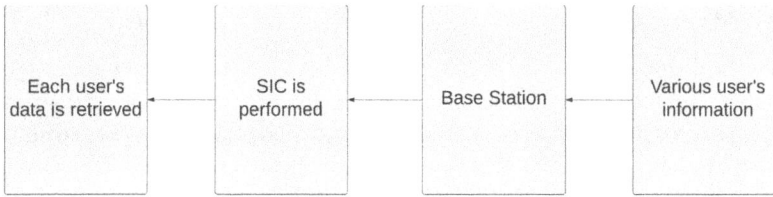

FIGURE 5.3 Uplink NOMA transmission.

5.3.2 UPLINK NOMA TRANSMISSION

According to Figure 5.3, each cellular user broadcasts their signal to the BS in an uplink NOMA channel. In order to identify the signals of cell phone users at the ground station, SIC loops are run. BS broadcasts power values to mobile users and that the uplink and downlink channels are mutually exclusive [21]. On the basis of the places, they are as in the downstream, the UEs could further optimize their transmitting intensity in the upstream. In this specific scenario, we presume that consumers are equally dispersed over the cell coverage area and receiving energy levels from separate consumers. From a practical standpoint, this assumption makes more sense because power optimization calls for connections between all User equipments (UEs), which may be challenging to achieve. The initial code that recognizes will be the message sent by the nearby customer [20].

5.4 ADVANTAGES OF NOMA

5.4.1 IMPROVED SE

Compared to different techniques, NOMA offers higher SE since several users can access the facilities over each resource block (RB), but in OMA, only one RB is allotted to each user, resulting in bandwidth loss [22]. In order to increase the network's throughput, NOMA may be simply combined with other 5G techniques including massive multiple-input multiple-output (mMIMO), device to device (D2D), mm-wave, cognitive radio (CR), as well as the heterogeneous networks (HetNets). Both in the power domain as well as the code domain of NOMA, users share the time-frequency source non-orthogonally. As mentioned earlier, although the upstream of additive white Gaussian noise (AWGN) structures can achieve the largest possible cumulative capacity thanks to NOMA, which also makes consumer fairness more logical.

5.4.2 IMPROVED THROUGHPUT

It is observed that NOMA in 5G achieves better throughput compared to other techniques. In the downstream of AWGN networks, NOMA's capacity bound is greater than OMA's. OMA can achieve the largest sum capacities under conditions of inter-symbol interference (ISI) in the downstream of multiple pathways' path loss, whereas NOMA relying on MUD is preferable.

5.4.3 Massive Connectivity

The non-orthogonal properties of NOMA enable it to host billions of mobile devices. Given that the packets are smaller and more irregular in nature, it is appropriate for both the Internet of Things (IoT) [23] as well as the internet [24]. Whereas in OMA one device purchases one RB, RBs are wasted; in NOMA numerous devices receive the functions through one RB. The allocation of non-orthogonal resources in 5G NOMA shows that the number of sustainable users or devices is not necessarily constrained by the quantity of available orthogonal assets. NOMA can greatly boost the amount of simultaneous connections, potentially enabling huge connectivity. Of course, it should be highlighted that various concerns with NOMA systems' practical implementation, such as their hardware flaws as well as computational complexity, may prevent the realization of vast interconnection (Figure 5.4).

5.4.4 Low Latency

The HetNet architecture used in 5G necessitates stricter latency constraints. Data transmission in LTE requires an access permission request that takes 15.5 ms [25]. This makes it difficult to easily satisfy the radical criterion of keeping a user latency below 1 ms [26]. With some uplink NOMA methods, however, dynamic scheduling is indeed not necessary. NOMA, which provides grant-free transmission, is utilized to solve the problem, particularly in the uplink (UL) situation. Moreover, NOMA offers flexible scheduling across a large number of devices in accordance with application requirements and device quality of service (QoS).

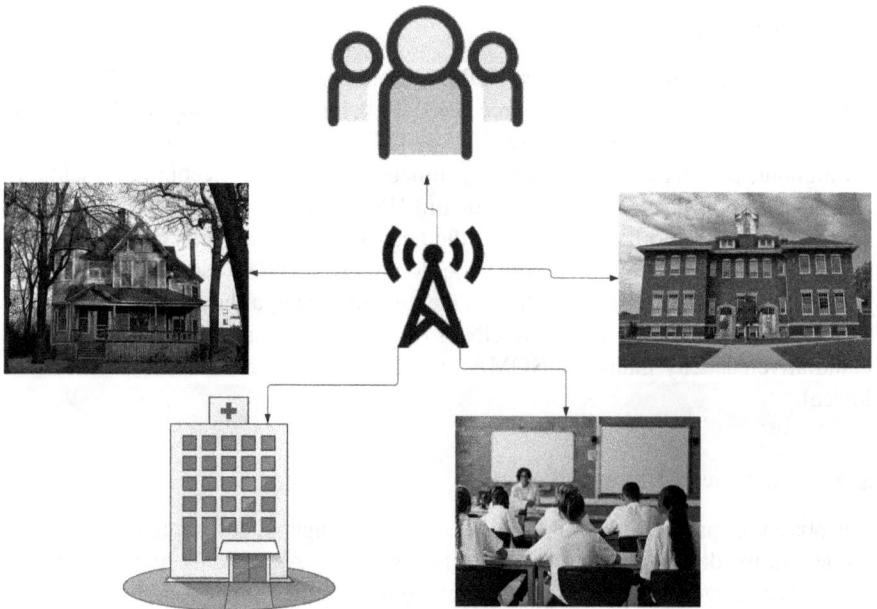

FIGURE 5.4   Massive connectivity in NOMA.

5.4.5 Low Signaling Expense

Grant-free uplink transmission in NOMA can be advantageously implemented and both the transmission delay and the signaling overhead can be drastically decreased. Take note that the SIC procedure may add additional latency to some NOMA systems using SIC receivers. In order to serve more users, enhanced MIMO methods can be used; however, the quantity of users dependent on SIC should not be high [27].

5.4.6 Relaxed Channel Feedback

As CSI (channel state information) feedback is solely used mostly for power distribution in power-domain NOMA, the channel feedback requirement will also be relaxed. Therefore, exact, immediate CSI information is not required. Therefore, regardless of if static or portable users are enabled, delivering outdated, less-accurate network feedback tied to a specific maximal imperfection and lag is unlikely to significantly affect the feasible system performance unless the channel fluctuates tremendously.

5.4.7 Fairness

Because NOMA ensures user equality, more power is distributed to weak users (those with bad channel quality) and less to dominant users. Then, QoS in regard to data rate is assured for both dominant as well as weak users. The researchers in [28] offer fair power distribution techniques to improve fairness among various users. Cooperative multipoint transmission (CoMP) and cooperative transmission are also essential for improving the fairness of the QoS for the weaker users [29, 30]. Also, a power-allocation strategy is suggested in [31] considering average CSI there at the transmission in order to increase user fairness.

5.5 NOMA SOLUTIONS

5.5.1 Cooperative NOMA

Cooperative transmissions have drawn a lot of attention in wireless connections because they can provide spatial variety to reduce fading while eliminating the challenges of placing many antennas on compact communications terminals [32]. Many relay nodes are allocated to help a source forward data to the appropriate recipients in cooperative transmissions. Consequently, integrating cooperative transmissions with NOMA 5G can enhance the network's capacity and reliability [30]. According to this plan, consumers with good channel conditions interpret data for the others, serving as relays to increase the dependability of message reception for users with weak links to the ground station. In [30], a cooperative NOMA mechanism is suggested. Similar to the standard NOMA 5G, C-NOMA also detects the multiuser message using a SIC receiver. In order to increase the reception dependability of the consumers experiencing bad channel conditions, it is possible to rely on the users linked with superior channel conditions as relays. The main distinction between cooperative transmission and the fundamental NOMA related with SIC is the ability to deliver signals from customers benefiting from good channel conditions to customers with bad channel conditions using short-range communication skills, such as Bluetooth as well as ultra-wideband mechanisms.

5.5.2 Multiple Access with Successive Interference Elimination

Let's take a look at a K > N, or rank-deficient, upstream Successive interference cancelation amenable multiple access (SAMA) network that supports K consumers with the help of N equilateral orthogonal frequency-division multiple access (OFDMA) subcarriers. The spreading grid B = (b1; b2; bK) is built based on the following principles [33], and the framework of SAMA is identical to that used by MUSA. However, in SAMA, the non-zero components of any spreading series bK for customer k are equal to one.

- The maximum quantity of groups in the spreading series should have varied numbers of ones.
- It is best to have the fewest possible overlapped spreading patterns with the same quantity of ones.

The message forwarding method is used at the receiver to separate the signals from the various users. Efficient interference cancellation is the design goal pursued in establishing the spreading grids in SAMA [34]. Since the first user's spreading series contains four non-zero items, the associated diversity order equals four. The most trustworthy symbol is that of the first user. Because of this, the symbol of the first user may be quickly ascertained in a few repetitions, which is advantageous for the resolution of a symbol detection mechanism of all other consumers with lesser diversity orders.

5.5.3 Energy-Efficient NOMA

NOMA in 5G can achieve better SE. Although SE demonstrates the effective use of a finite spectrum resource, it offers no insight into the effective use of energy. Because of the current spike in curiosity about greener telecommunications, reducing energy utilization has been of highest importance to experts as well as 5G also has emphasized energy efficiency (EE) as among the basic necessities to be fulfilled. Yet, Shannon's theorem [35] shows that it is not possible to maximize SE while also decreasing energy usage, requiring a tradeoff. It should be noted that the EE against SE graph always has an ideal point when switch power is taken into account. EE single unit NOMA with two users was researched in [36]. The EE-SE connection can be discovered to be straight with fixed total energy usage. Any element along the EE-SE graph is possible with the proper power distribution among two users. Maximum EE efficiency can be attained for a specific SE for every user. Using power-control techniques, the total energy can be changed to alter the level of effectiveness [37].

5.5.4 NOMA in CoMP

In multi-cell scenarios, NOMA 5G also shows its own advantages, which inspired the idea of NOMA with CoMP. However, NOMA in case CoMP may cause significant inter-cell interference if a single-cell NOMA layout is used directly to multi-cell settings. The signals of all the NOMA users can be used in cooperative transmit-precoding to reduce inter-cell disturbance in the downlink. Finding the best transmit-precoder is not simple, and all users' information and channel statistics should be present at the BSs concerned. Furthermore, a beam produced by geographically distant BS antennas might not be able to effectively cover greater than one angular displacement user for intrabeam NOMA, making multiuser broadcast precoding of solo-cell NOMA in a CoMP situation ineffective. A reduced-complexity transmission precoding approach for NOMA 5G in CoMP was developed by taking use of the fact that various users' carrier-to-interference ratio (CIR) are probably going to be somewhat varied in the multi-cell situation [38]. The rate coverage likelihood of a user with rank m (in regard to its distance from its own feeding BS) among all consumers within a cell as well as the mean rate coverage likelihood of all consumers within cell have been analyzed using the concept of order derivatives and the poisson cluster procedure. Likewise, a multi-cell upstream NOMA scheme has been taken into consideration in [39]. It has already been demonstrated that a NOMA cluster's mean rate coverage is superior to its corresponding OMA cluster. Aside from SIC implementation challenges flawed CSI, multiuser power distribution, and clustering, [40] also highlights the key practical difficulties that are faced in implementation of multiple cell NOMA.

5.5.5 Impact of Path Loss

NOMA may serve as a viable power-domain client multiplexing strategy for future wireless systems, according to a survey by [41]. The path loss (PL) effectiveness of NOMA users can be examined in two different instances. Every consumer has a specified bit speed in the primary situation that is decided through the assigned service quality. Since it evaluates NOMA's capacity to satisfy users' QoS needs, the path loss is the optimal performance indicator in this case. To assess NOMA functionality in this case, the attainable ergodic total rate can be looked into. If users' bit rate and allocated power are properly selected, NOMA can provide superior outage evaluation than other OMA systems, claims Ding et al. [42]. The results of this study also demonstrated NOMA's better ergodic total rate. Inter-user disturbance also happens in NOMA. Hence, the achievable benefit will always be constrained by the rank response in NOMA. To get over this restriction, a few rank optimization techniques [43] could be used, improving the performance of the ergodic capacities as well as outage probability.

5.6 PERFORMANCE EVALUATION OF NOMA

5.6.1 Throughput

A key parameter that shows the productivity of a ground station that has been used is the bit rate of a user. The NOMA models that function in a delay-limited transmits channel make use of the throughput estimated scenarios. When SIC is successful, customer throughput is better than when SIC is unsuccessful. Many approaches, including [44, 45], can be used to tackle the throughput optimization problem. However, when building NOMA systems, it is also necessary to take into account the tradeoff among the throughput as well as other criteria, such as fairness, sensing, etc. [46–48].

5.6.2 ENERGY EFFICIENCY (EE)

One of the important characteristics that have shown that NOMA technology works better than the existing technologies is enhanced energy efficiency (EE). A key performance parameter for new wireless systems is EE [49]. The sum of the feasible data rates is divided by the total power utilized by the entire network to determine the EE. Based on the amount of consumers and the floating energy used by each unmanned aerial vehicle (UAV), the energy utilization of UAV structures may be evaluated [50]. Consequently, in energy harvesting allowed NOMA systems, the EE-maximizing challenge is a critical goal that must be attained. There are a number of procedures that can be used to handle the best EE challenges, including [51–53]. NOMA 5G scheme EE obtains a greater value at a lower SNR in several documented research efforts [54, 55]. In contrast, when the SNR increases, this EE falls to zero. EE is important because information and communication technology (ICT) account for 15 percent of global energy consumption, which is a major problem for both society and the human race as a whole. In down link-NOMA (DL-NOMA), distribution is carried out by focusing on EE and assuming bits/joule to assess EE success. By isolating the main issues in subchannel allocation and power assigning under an EE-optimization limitation, the issues have been articulated. Here, single subchannels can multiplex a maximum of two users, making SIC calculations simple [56].

5.6.3 SPECTRAL EFFICIENCY (SE)

In both UL as well as DL broadcasts, NOMA is the most effective spectrum-use strategy. This is because each NOMA user utilizes the entire bandwidth, as opposed to OMA users who are limited to using only a portion of the bandwidth. Moreover, NOMA is well suited to work over a wider time or space span in conjunction with other techniques, like MIMO, cluster, as well as mm-wave, to achieve even higher throughput [56]. NOMA has received particular attention from the domains of education, innovation, research, as well as industry as a result of the benefits listed.

5.6.4 OUTAGE PROBABILITY

The possibility that the immediate attainable data rate is less than the intended rate is referred to as the outage probability and is used to assess a NOMA information system. The NOMA outage probability equations are derivable from closed-form formulae and can be tested using numerical computations. Both the standard signal noise ratio (SNR) regime as well as the higher SNR regime can be used to analyze the outage likelihood. The NOMA-based modeling methods that were researched over the past ten years nearly achieved higher outage likelihood than the traditional

OMA [44]. In NOMA 5G systems, SIC-based decoding can lead to faulty SIC. Hence the likelihood of a system outage is also impacted. At a higher SNR situation, the outage probability provided by [57] is essentially true.

5.7   NOMA CHALLENGES

Researchers today have worked on developing and implementing NOMA systems along with solving several problems associated with these approaches. However, there are numerous challenges and open issues that need to be addressed in 5G NOMA techniques. This section attempts to point out some research directions for researchers who wish to investigate NOMA on a larger scale. Hence the major challenges in NOMA for 5G networks are discussed.

5.7.1   IMPACT OF TRANSMISSION DISTORTION

The transfer of source data via communication channels, such as video and audio, is generally termed as lossy transmission. While travelling to the receiver, the sent data are always distorted. To date, a lot of theoretical thought has gone toward determining source quality over fading channels in order to address this lossy transmission. To reduce end-to-end distortion, several source coding as well as channel coding have been framed. Unfortunately, the desirable levels of distortion, expense, and complexity are in conflict when source coding and channel code diversity are used. Choudhury and Gibson evaluated the source distortion for the ergodic and outage definitions of channel capacity [58]. Data capacities along with distortion are both impacted by the possibility of an outage. It is apparent that an outage possibility that maximizes outage occurrence would not provide the smallest percentage of projected damage [59].

5.7.2   RESOURCE ALLOCATION

In order to meet a range of traffic demands, 5G networks offer extremely high data throughput at incredibly low lag in reliable ways. Due to restricted resources, this is a highly challenging task. So, resource management must support efficient consumption. When and quantity of associated assets that should be supplied to each individual must be determined through a set of procedures known as wireless asset management [60]. Also, it is based on the kind of resources. bandwidth (BW) belongs to these cellular resources, in accordance with Shannon's theorem. The overall BW is initially divided into numerous sections as part of the efficient administration of the overall BW in a transmission medium. Each piece is then distributed to a particular user or group of users, much like NOMA. Additionally, the overall number of packets sent by each user varies as time passes. For customer pairing as well as effective distribution of power among consumers in 5G NOMA, an intricate approach is consequently required so as to get the greatest outcomes while expending a minimum of resources. The distribution of resources in 5G NOMA may be investigated from the perspective of mathematical optimization methodology. In [61], Lei et al. took into account combined power and route allotment with NOMA in 5G technology. To address the power and route distribution issue, they consider user power control as well as SIC implementation. The distribution of power in cognitive networks based on NOMA is another undiscovered region that needs study [62].

5.7.3 OUTAGE PROBABILITY ANALYSIS

Understanding the functionality of many wireless systems begins with outage analysis. In reality, user outage conduct determines the possible capacity. Several scholars have generally looked into the likelihood of an outage in the fundamental NOMA architecture. For instance, when users are randomly distributed in a cell, NOMA experiences less outages than OMA [41]. Path loss was taken into account in that analysis. Moreover, a NOMA-BF system was found to have a higher sum capacity than a traditional multiuser BF network [63]. In a similar manner, [36] the primary aim of NOMA is interference prevention between cells through precoder design. However, extra study is needed in order to totally comprehend the manner in which cell-edge consumers act when there are disruptions. Additional study can be conducted for assessing outage productivity. In this study, NOMA has been applied to interior VLC (visible light communication) networks based on actual interior channel circumstances [64] and NOMA in 5G with coding [65, 66].

5.7.4 CONSISTENT FAIRNESS

A large number of venues in mm Wave portable have a transmission breakdown at an altitude that exceeds 175 m [67]. The real disruption might be severe and substantial when there are additional regional barriers because outage is strongly dependent on the surroundings. It would be beneficial to create a NOMA method that provides customers with consistent outage encounters, particularly for individuals who reside far away.

5.7.5 DYNAMIC USER PAIRING

Co-channel interference in 5G NOMA systems is very strong as it shares the same spreading code, time, and frequency with multiple users. Therefore, it is hard to request all consumers of the case to execute NOMA together. Instead, those who utilize the system may be split up into various groups, each of which would make utilization of NOMA. Orthogonal bandwidth resources are allotted to the various groups. Normally, the still situation is taken into consideration where the xth client along with the yth client is paired to execute NOMA in 5G systems. To maximize NOMA's benefits, dynamic user pairing or grouping schemes must be designed, despite their difficulty in practice. In this context, the analytical findings obtained [68] in the case of two selected users are taken into account. It is well acknowledged that this UPPA (user pair power allocation) technique [69] enhances the performance gains.

However, NOMA technology can only set up two clients at a time. A promising expansion of the UPPA technique can be an energy distribution technique in case of several users in a NOMA system where multiple users are allowed by the scheduler. For applying UPPA technique [70] to a more practical system which has a coding scheme as well as a suitable modulation, more research is needed to be performed.

5.7.6 IMPACT OF INTERFERENCE

Although interference analysis is a general term in the field of wireless communications, this survey concentrates on cooperative NOMA [30], which uses Bluetooth interactions in this mode. Unfortunately, usage of Bluetooth radio for mobile interactions is severely hampered by the prevailing WPAN (wireless personal area network) processes. The customer's discovery phase is hampered by BT interference as a result of the diminished range along with speed; intermittent as well as permanent link reduction add difficulty in pairing. In reality, the distribution of channel allocation is influenced by interruption in the deployed domain, payload size, as well as length across cooperative users. The customers in 5G NOMA are matched as per CSI, hence a scatternet that organizes itself to manage BT connections requires to be redesigned to work with 5G NOMA. A strong scatternet should also provide high probability valid paths between nodes even when certain wireless links are completely lost due to user movement. Additionally, interference changes in a dynamic way as a result of user mobility.

5.7.7 PRACTICAL CHANNEL MODEL

Next-generation cellular systems need both spectrum availability and a reliable radio access method to handle the ever increasing amount of user data. It is currently clear that 5G will make use of the available spectrum in the unoccupied millimeter-wave (mmW) wavelength ranges. Moreover, mmW wireless connections are anticipated to replace copper and fiber in the backbone networks for 5G, enabling quick installation and mesh-like connections. The mmW spectrums between 30–300 GHz, which indicate a new market for cellular systems, offer a significant quantity of bandwidth Understanding the challenges of mmW transmissions and route dynamics is the key. For the advancement of 5G cellular networks this is a requirement [67]. A greater degree of precision would be displayed if it were possible to portray the radio network within the mmW region using the recorded route loss and postpone spreading facts [71].

5.7.8 HETEROGENEOUS NETWORKS (HETNET)

For 5G cell phone networks, the HetNet provides adequate bandwidth, reach, and low usage of energy. In comparison to small density distribution of fewer high-energy units, infrastructure with a large density distribution of limited power units can also greatly boost EE. Many studies on HetNets have been conducted, including those on node cooperation, ideal load balancing, and improved inter-cell disturbance coordination [72].

Recently, [73] examined the system architecture of such a cooperative HetNet system framework for 5G with the objectives of both EE as well as spectrum efficiency. NOMA unique usage in specific HetNets can provide additional profit because the goal of NOMA and the HetNet are the same. The uneven spatial geographical distribution of consumers of cell phones will also have an effect on how NOMA performs.

5.7.9   NOMA with Multiple Antennas

The quantity of de-correlated pathways obtained increases with the degree of the MIMO channel grid, improving system performance. Consequently, while assessing MIMO NOMA in 5G, the network matrix rank is crucial. Existing MIMO NOMA research, however, takes full-rank network matrices into account when analyzing system efficiency. The analytical outcomes produced by Ding et al. [74] for this constraint, for instance, give the top limits of outage probability as well as capacity. To upgrade MIMO NOMA functionality, rank modification for NOMA may be implemented [43]. Focusing on current MIMO NOMA approaches in conjunction with other intriguing wireless technologies, such as orthogonal frequency as well as code division multiplexing, is equally intriguing (OFCDM). The majority of the MIMO NOMA 5G algorithms examined in this study have large computational complexity, such as the beamforming technique presented by Kim et al. [75].

5.7.10   Carrier Aggregation

LTE-A makes use of the carrier aggregation (CA) idea to boost the BW as well as the bit rate of its customers [76]. According to the CA approach, users receive shared resources made up of two or even greater components. Every carrier in the aggregate makes up a component carrier (CC). Contiguous distribution, where CCs are placed next to one another, or non-contiguous distribution, where there is a space in between, would be the two possible arrangements for aggregation. It is feasible to combine CA and NOMA in 5G to benefit from their respective advantages. To do this, a different user pairing is going to be utilized than in standard NOMA. It is commonly known that in the basic two-user NOMA design, a NOMA subscriber is paired with some other NOMA user; typically, each carrier is allocated to two users depending on channel circumstances. The following justifies the customer paring in CA-enabled NOMA. The CC of one NOMA user will be the CC of some other NOMA consumer. The user is connected with regard to the CC in this instance. In addition, it is also feasible that the user and regard to the CC are paired. If CA is linked with basic NOMA, a NOMA user may be paired with several distinct users simultaneously based on the quantity of CCs. The problem of which CA type is ideal for NOMA solutions is still unresolved; making an analysis of several CA types in 5G NOMA an intriguing research topic.

5.7.11   NOMA with Antenna Selection

There must be an equivalent count of concurrent RF links at the entrance side to accommodate the utilization of multiple antennas simultaneously on the transmitter side. As a result, adding more antennas results in an increase in system complexity, power use, and expense. Transmit antenna selection (TAS), which simplifies the transmitter structure, is frequently preferred to address such problems. The practical TAS approach only uses one RF link for output and arbitrarily picks the best antenna among a wide range of antennas. Finally, this leads to a decrease in complexity of systems, price, dimension, and energy use at the sacrifice of a bearable performance decline. Research can discover a unique TAS-NOMA method for downstream transmissions from a ground station outfitted with numerous antenna to several consumers, each outfitted with only solo antennas. When you take into account that the intended consumer rate is distributed strategically according to network situation, the scheme's ultimate goal should be to increase the sum rate.

5.7.12 Other Challenges

Few additional issues that must be resolved in NOMA are discussed here. For instance, in a downstream situation, the sender distributes the energy to the clients according to their own unique CSI. For obtaining robust performance, a good CSI feedback system, channel estimation strategy, and reference signal architecture are crucial. Moreover, NOMA's benefits can only be seen under the optimum circumstances of flawless CSI data capture at the transmission end. Using a restricted feedback channel to obtain CSI is one potential fix [77]. Yet, in order to communicate different channel quality indications, more BW is needed. According to this viewpoint, proper user selecting and power distribution strategies are required. peak-to-average power ratio (PAPR) may cause the sender's power to operate in an irregular operational area in multicarrier transmissions. The resultant signal distortion in the PA output is considerable. It is critical to take PAPR's effects into perspective when choosing which strategies to employ for the optimal NOMA performance. To increase capacity, non-orthogonal synchronized transmissions between a number of small cells as well as a macro cell can use the relaying [78] idea to extend cell coverage. To improve performance in terms of sum rate, Kalokidou et al. [79] suggest a hybrid scheme that incorporates the ideas of topological interference control and NOMA methods. It is possible to conduct additional research on this hybrid method's MIMO variant. Yet, choosing a proper power allocation strategy is difficult, particularly for a dense system. Despite the fact that Xiong et al. proposed a software defined radio (SDR) based NOMA-like a model [80], further over-the-air tests are necessary to show NOMA's viability in 5G mobile services.

5.8 APPLICATION OF NOMA

5.8.1 In Case of Cellular Networks

The NOMA implementation can be applied to both uplink as well as downlink transmissions in cellular systems. In uplink imparting, NOMA 5G is used at the BS, and the SC process sums up signals received from users who broadcast at various power levels. In contrast, during downlink imparting, the supplier provides users a mixed signal with varying levels of power for every individual signal, and the receiver uses the SIC approach to recognize and decode its individual signal. In comparison to earlier cellular network generations, such as 1G to 4G, NOMA can offer improved spectrum efficiency but also user fairness in 5G. Moreover, MIMO NOMA, which makes use of multiple antennas, can be used to benefit from NOMA. The goal of the numerous antenna techniques is twofold.

In the initial case, beamforming is created, which lowers overall signal-to-interference-plus-noise ratio [81]. The other involves building spatial multiplex, which improves throughput [82]. By boosting the signal-to-interference plus noise ratio (SINR), the spectral efficiency of the beamforming NOMA is enhanced. Whereas NOMA, along with spatial multiplexing, may increase its gain by employing numerous antennas. One transmit antenna sends one independent data stream, hence the possible rate with this kind of NOMA can be increased in proportion to the number of transmit antennas. Communication between the BS and customers with the use of dedicated relay is another utilization of NOMA.

The following four groups can be used to categorize cellular networks (CNs). There are four types of CNs: single-cell single-tier (SCST), single-cell multi-tier (SCMT), multi-cell single-tier (MCST), and multi-cell multi-tier (MCMT). SCST CNs are the most prevalent sort of NOMA-based CN, which have been thoroughly studied in recent research. The NOMA approach with SIC receivers was first conceptualized in the early works of [83–85], who then confirmed their idea through system-level simulations. The authors found that NOMA achieved a significant performance boost of 30–35 percent. A pioneering work that examined the performance of a single-cell downstream NOMA network with randomly placed users was of great help to researchers [42].

5.8.2 In UAV Networks

Unmanned aerial vehicles (UAVs) are currently proving their value in civil applications like aerial photography, improved freight distribution, and the management of wildfires as well as natural disasters [86, 87]. A possible use for UAV-involved communication has emerged in the business and academic worlds. Multiple access strategy has emerged as a crucial element for integrating UAV in 5G as well as future networks. Due to its impressive qualities, such as better spectral efficiency, low latency, and high connectivity, NOMA is one of the key choices in these systems [88]. By implementing PD NOMA to assist people who use the very same frequency/time asset, NOMA can support the high degree of connection in UAV networks. Hence, there are no obvious channel gain differences between UAV and base users due to line of sight (LoS) connectivity. Because of this, NOMA-assisted UAV systems need to be modified to take into consideration significant fading differences across NOMA participants, like user pairing strategies. Aerial users (AUs) and terrestrial users (TUs) frequently have high data needs during uplink, some of which may be latency-sensitive. In NOMA, the powerful user receives all of the power distributed. Yet, NOMA emerges as a superior solution when the RBs are unable to service all consumers orthogonally. The tradeoff among system performance, delay, and user fairness can be specifically balanced via NOMA. As a result, multiple users may consistently meet latency requirements while ensuring great spectral efficiency. Moreover, AUs typically exhibit higher macro-diversity. To improve performance, it may be possible to make use of the significant macro-diversity [89].

5.8.3 MIMO NOMA

MIMO-NOMA architecture is exceedingly complex compared to single-input-single-output (SISO) NOMA, primarily for the two reasons listed below. So first of all, even though MIMO NOMA performs better than MIMO OMA, it is unclear if using it will result in the best system effectiveness [90]. In SISO, it is evident that the employment of NOMA would achieve a portion of the transmission channel's portion when using downstream NOMA. The likelihood that NOMA will obtain maximum individual and sum rates is carefully determined in [91]. Hence, it is difficult to evaluate MIMO-NOMA performance. MIMO-NOMA utilization can produce similar productivity as DPC (dirty paper coding) if the users' channels are quasi-degraded, as demonstrated in [92] (i.e., implementation of NOMA yields the best result in the MIMO scenario).

5.8.4 Millimeter-Wave NOMA

One of the important key technologies for 5G had been recognized as mmW transmission [93]. The restricted amount of spectrum resources less than 6 GHz that is accessible for wireless communications which drives both mmW communications and NOMA. In contrast to NOMA, which improves bandwidth efficiency, mmW makes use of the less-used mmW frequency ranges. A significant step forward for mmW communications came when the FCC (Federal Communication Commission) authorized the release of larger than 10 GHz of bandwidth for 5G communication systems [94]. The adoption of NOMA is crucial despite the abundance of available spectrum in the mmW range for two main considerations. So first of all, NOMA 5G usage offers a crucial tool for supporting large connection. Second, the benefit from employing the mmW bands will be swiftly eclipsed by the explosive development in demand for new data services. According to [95], an average throughput of 1000 Gbps data is needed to give a good quality telepresence, for instance. NOMA can be used to efficiently increase mmW communications' spectral efficiency and meet their fast-expanding demand (Figure 5.5) [96].

5.8.5 Device-to-Device Connectivity Using NOMA

A cooperative NOMA network incorporating full-duplex (FD) capabilities was taken into consideration by the authors in [97]. To service both powerful and powerless customers, the BS uses the NOMA 5G protocol. The authors suggested using FD at the powerful user to improve the weaker NOMA user's outage performance. In this configuration, users interact with one another using cooperative D2D transmission that is OMA-based. The authors of [33] also took into account a cellular system with D2D communications as an underlay, wherein D2D consumers can also employ FD mode. The authors suggested a novel-specific condition for choosing among D2D versus FD modes. The performance of the network under consideration is assessed by developing closed-form outage probability equations, which are dependent on stochastic geometry approaches. One hybrid gateway is employed by [98] to acquire from both D2D as well as cellular users in a linked D2D transmission that underlies

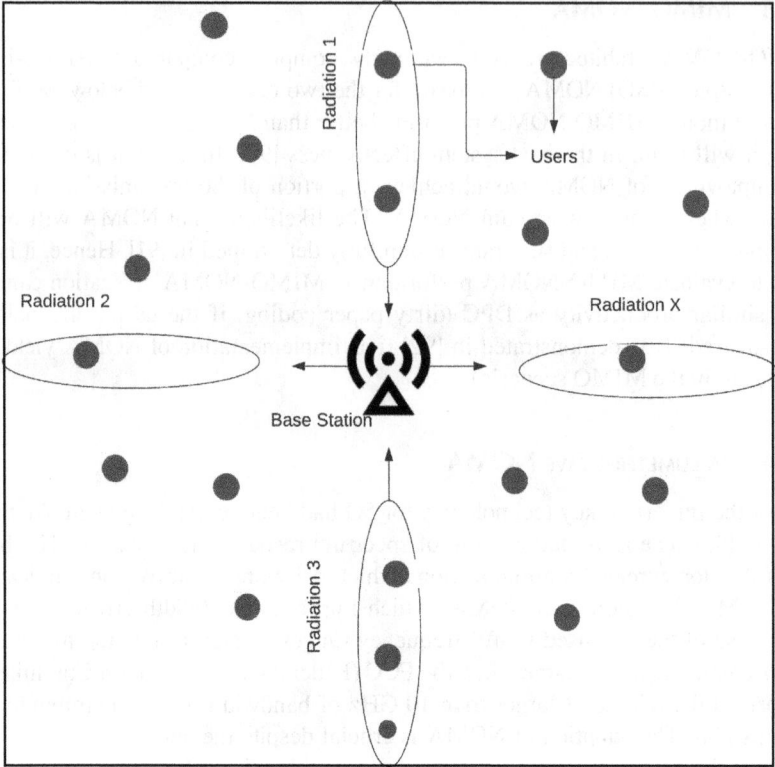

FIGURE 5.5 mmWave-NOMA communication system.

a NOMA-based mobile network. They specifically go into the issue of allocating resources and present a D2D pair low complexity, fuel efficient method that satisfies cellular customers' QoS needs. The results showed that their suggested methodology quickly arrived at the best option while still using less energy than the alternatives. In [99], the idea of NOMA-assisted D2D relaying is put up, where the broadcast consists of two phases. The BS broadcasts the NOMA signal to nearby and distant users during the first phase. By taking advantage of the next D2D node's ability to overhear this NOMA communication, this relay node broadcasts both its own signal and overheard NOMA information from the first part to the D2D receiver in the next part. This improves the performance of distant users. The D2D relay node employs NOMA, although it still engages in paired D2D transmission with a sole D2D receiver.

5.8.6 Wireless Sensor Systems with NOMA

The investigation of the use of NOMA with wireless sensor networks (WSN) in the existing literature is limited. The research in [100] is recognized as a pioneering study of NOMA-based WSNs. The authors presented and modelled a NOMA-based WSN wherein detectors and sink units are assumed to be freely dispersed

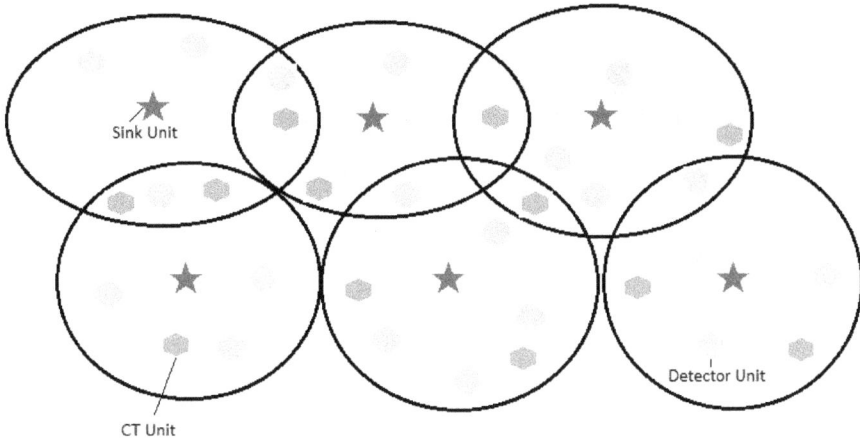

FIGURE 5.6 Wireless sensor systems with NOMA.

over the entire plane using stochastic geometry techniques. An energy-harvesting, two-hop NOMA-based WSN was taken into consideration by the authors in [100]. The authors suggested an energy harvesting methodology based on two different sorts of depending. Closed-form equations for the path loss and ergodic frequency are constructed in order to assess the performance. The proposed strategy is able to achieve a much reduced outage probability as well as better rates when compared with previous strategies, according to numerical data. Moreover, a WSN based on NOMA and utilizing a time changing energy harvesting mechanism is discussed in [102]. Outage probability and attainable rate analytical closed-form formulas are developed. The simulations are run in order to confirm that the results are accurate. [103] investigates the use of NOMA with WSN for intelligent farming system. The performance of the uplink NOMA network was taken into consideration by the authors, who derived closed-form analytical formulas for the average sum rate as well as outage likelihood. According to the simulation outcome, the suggested relay-aided NOMA-based method outperforms the traditional approach for WSN in farming (Figure 5.6).

5.9 CONCLUSION

This chapter provides a detailed overview of the NOMA technology in 5G. It covers NOMA and its two main types, P-NOMA and C-NOMA. Workflow includes NOMA transmission at the downlink and uplink respectively. Various NOMA advantages and potential solutions for NOMA in 5G have also been discussed. It briefly covers the performance evaluation of various NOMA 5G parameters such as user data rate, capacity, SE, EE, etc. In addition, this chapter also focuses on several challenges faced by NOMA technique in 5G. Lastly, it also discusses various NOMA 5G applications in different fields. Hence, this chapter is expected to provide a detailed understanding to all individuals who explore NOMA 5G wireless technologies.

REFERENCES

1. A. F. M. Shahen Shah, A. N. Qasim, M. L. Karabulut, H. Ilhan, and MD. B. Islam, "Survey and performance evaluation of multiple access schemes for next-generation wireless communication systems" *IEEE Access*, August 2021. DOI: *10.1109/ACCESS.2021.3104509*

2. V. Basnayake, D.N.K. Jayakody, H. Mabed, A. Kumar, and T.D.P. Perera, "M-Ary QAM asynchronous-NOMA D2D network with cyclic triangular-SIC decoding scheme" *IEEE Access*, January 2023. DOI: *10.1109/ACCESS.2023.3236966*

3. E. Hwang, X. Hao, and G.E. Atkin, "Design of incidence matrices with limited constellation expansion in massive connectivity NOMA systems" *IEEE Access*, January 2023. DOI: *10.1109/ACCESS.2023.3235825*

4. N. Jaiswal, A. Pandey, S. Yadav, N. Purohit, and D.S. Gurjar, "Physical layer security performance of NOMA-aided vehicular communications over Nakagami-*m* time-selective fading channels with channel estimation errors" *IEEE Open J Veh. Technol.*, December 2022. DOI: *10.1109/OJVT.2022.3222187*

5. S. Chege, T. Walingo, and F. Derraz, "Codebook design for PD-SCMA NOMA systems" *IEEE Access*, February 2023. DOI: *10.1109/ACCESS.2023.3242697*

6. R. Zhu, J. Guo, F. Wang, B. Lin, and Y. Chen, "Spectrum efficient resource allocation of NOMA downlink system with MMSE receiver" *IEEE Access*, February 2023. DOI: *10.1109/ACCESS.2023.3258451*

7. Y.Y. Guo, X.-L. Tan, Y. Gao, J. Yang, and Z.-C. Rui, "A deep reinforcement approach for energy efficient resource assignment in cooperative NOMA enhanced cellular networks" *IEEE Internet of Things J.*, 2023. DOI: 10.1109/JIOT.2023.3253129

8. A. Benjebbour, Y. Saito, Y. Kishiyama, A. Li, A. Harada, and T. Nakamura, "Concept and practical considerations of non-orthogonal multiple access (NOMA) for future radio access," in *Proc. Int. Symp. Intell. Signal Process. Commun. Syst. (IEEE ISPACS)*, Naha, Japan, Nov. 2013, pp. 770–774.

9. Y. Saito, Y. Kishiyama, A. Benjebbour, T. Nakamura, A. Li, and K. Higuchi, "Non-orthogonal multiple access (NOMA) for cellular future radio access," in *Proc. IEEE 77th Veh. Technol. Conf. (VTC Spring)*, Jun. 2013, pp. 1–5.

10. K. Higuchi and Y. Kishiyama, "Non-orthogonal access with random beamforming and intra-beam SIC for cellular MIMO downlink," in *Proc. IEEE Veh. Technol. Conf. (VTC Fall)*, Las Vegas, NV, USA, Sep. 2013, pp. 1–5.

11. "TP for classification of MUST schemes," 3GPP, Sophia Antipolis, France, Tech. Rep. R1-154999, Aug. 2015.

12. *Downlink Multiuser Superposition Transmissions for LTE*, document RP-160680, 3GPP, Sophia Antipolis, France, Mar. 2016.

13. M. Zeng, A. Yadav, O. A. Dobre, G. I. Tsiropoulos, and H. V. Poor, "Capacity comparison between MIMO-NOMA and MIMO-OMA with multiple users in a cluster," *IEEE J. Sel. Areas Commun.*, vol. 35, no. 10, pp. 2413–2424, Oct. 2017.

14. L. Liu, J. Tong, and L. Ping, "Analysis and optimization of CDMA systems with chip-level interleavers," *IEEE J. Sel. Areas Commun.*, vol. 24, no. 1, pp. 141–150, Jan. 2006.

15. Z. Yuan, G. Yu, W. Li, Y. Yuan, X. Wang, and J. Xu, "Multi-user shared access for Internet of Things," in *Proc. IEEE 83rd Veh. Technol. Conf. (VTC Spring)*, May 2016, pp. 1–5.

16. H. Nikopour and H. Baligh, "Sparse code multiple access," in *Proc. IEEE 24th Int. Symp. Pers. Indoor Mobile Radio Commun. (IEEE PIMRC)*, London, U.K., Sep. 2013, pp. 332–336.

17. R. Hoshyar, F. P. Wathan, and R. Tafazolli, "Novel low-density signature for synchronous CDMA systems over AWGN channel," *IEEE Trans. Signal Process.*, vol. 56, no. 4, pp. 1616–1626, Apr. 2008.

18. J. Van De Beek and B. M. Popovic, "Multiple access with low-density signatures," in *Proc. IEEE Global Communications Conference (IEEE Globecom'09)*, Dec. 2009, pp. 1–6.
19. D. Guo and C.-C. Wang, "Multiuser detection of sparsely spread CDMA," *IEEE J. Sel. Areas Commun.*, vol. 26, no. 3, pp. 421–431, Apr. 2008.
20. R.C. Kizilirmak, "Non-Orthogonal Multiple Access (NOMA) for 5G Networks" *INTECH*, 2016. http://dx.doi.org/10.5772/66048
21. M. Aldababsa, M. Toka, S. Gökçeli, G.K. Kurt, and O. Kucur, "A tutorial on non-orthogonal multiple access for 5G and beyond" *Hindawi, Wirel. Commun. Mob. Comput.*, Article ID 9713450, 24 pages, 2018.
22. Y. Saito, Y. Kishiyama, A. Benjebbour, T. Nakamura, A. Li, and K. Higuchi, "Non-orthogonal multiple access (NOMA) for cellular future radio access," in *Proc. IEEE 77th Veh. Technol. Conf. (VTC Spring)*, Dresden, Germany, Jun. 2013, pp. 1–5.
23. M. Shirvanimoghaddam, M. Dohler, and S. J. Johnson, "Massive non-orthogonal multiple access for cellular IoT: Potentials and limitations," *IEEE Commun. Mag.*, vol. 55, no. 9, pp. 55–61, Sep. 2017.
24. M. Simsek, A. Aijaz, M. Dohler, J. Sachs, and G. Fettweis, "5G-enabled tactile Internet," *IEEE J. Sel. Areas Commun.*, vol. 34, no. 3, pp. 460–473, Mar. 2016.
25. L. Wang, X. Xu, Y. Wu, S. Xing, and Y. Chen, "Sparse code multiple access-towards massive connectivity and low latency 5G communications," *Telecommun. Netw. Technol.*, vol. 5, no. 5, p. 5, Jun. 2015.
26. "5G: Rethink mobile communications for 2020+," *FuTURE Mobile Communication Forum*, Nov. 2014.
27. A. Bayesteh, E. Yi, H. Nikopour, and H. Baligh, "Blind detection of SCMA for uplink grant-free multiple- access," in *Proc. IEEE Wireless Communications Systems (IEEE ISWCS'14)*, Aug. 2014, pp. 853–857.
28. S. Timotheou and I. Krikidis, "Fairness for non-orthogonal multiple access in 5G systems," *IEEE Signal Process. Lett.*, vol. 22, no. 10, pp. 1647–1651, Oct. 2015.
29. M. S. Ali, E. Hossain, and D. I. Kim, "Coordinated multipoint transmission in down-link multi-cell NOMA systems: Models and spectral efficiency performance," *IEEE Wireless Commun.*, vol. 25, no. 2, pp. 24–31, Apr. 2018.
30. Z. Ding, M. Peng, and H. V. Poor, "Cooperative non-orthogonal multiple access in 5G systems," *IEEE Commun. Lett.*, vol. 19, no. 8, pp. 1462–1465, Aug. 2015.
31. J. Cui, Z. Ding, and P. Fan, "A novel power allocation scheme under outage constraints in NOMA systems," *IEEE Signal Process. Lett.*, vol. 23, no. 9, pp. 1226–1230, Sep. 2016.
32. A. S. Ibrahim, A. K. Sadek, W. Su, and K. J. R. Liu, "Cooperative communications with relay-selection: When to cooperate and whom to cooperate with," *IEEE Trans. Wireless Commun.*, vol. 7, no. 7, Jul. 2008.
33. K.S. Ali, H. ElSawy, and M.S. Alouini, Modeling cellular networks with full-duplex D2D communication: A stochastic geometry approach. *IEEE Trans. Commun.*, vol. 64, pp. 4409–4424, 2016.
34. X. Dai, S. Chen, S. Sun, S. Kang, Y. Wang, Z. Shen, and J. Xu, "Successive interference cancelation amenable multiple access (SAMA) for future wireless communications," in *Proc. IEEE International Conference on Communication Systems (IEEE ICCS'14)*, Nov. 2014, pp. 1–5.
35. V. Ramachandran and A. Cheeran, "Evaluation of energy efficiency of spectrum sensing algorithm for Cognitive Radio networks," 2014 International Conference on Computer Communication and Informatics, Coimbatore, India, 2014, pp. 1–6.
36. S. Han, C. L. I, Z. Xu, and Q. Sun, "Energy efficiency and spectrum efficiency co-design: From NOMA to network NOMA," *IEEE Commun. Society MMTC E-Lett.*, vol. 9, no. 5, pp. 21–24, Sep. 2014.

37. Q. Sun, S. Han, I. C. Lin, and Z. Pan, "Energy efficiency optimization for fading MIMO non-orthogonal multiple access systems," in *Proc. IEEE Int. Conf. Commun. (ICC)*, London, Jun. 2015, pp. 2668–2673.

38. S. Han, I Chih-Lin, Z. Xu, and Q. Sun, "Energy efficiency and spectrum efficiency co-design: From NOMA to network NOMA," *IEEE MMTC E-Lett.*, vol. 9, no. 5, pp. 21–24, Sep. 2014.

39. H. Tabassum, E. Hossain, and M. J. Hossain, "Modeling and analysis of uplink non-orthogonal multiple access (NOMA) in large-scale cellular networks using poisson cluster processes," *IEEE Trans. Commun.*, vol. 65, no. 8, pp. 3555–3570, Aug. 2017.

40. W. Shin, M. Vaezi, B. Lee, D. J. Love, J. Lee, and H. V. Poor, "Nonorthogonal multiple access in multi-cell networks: Theory, performance, and practical challenges," *IEEE Commun. Mag.*, vol. 55, no. 10, pp. 176–183, Oct. 2017.

41. K. Higuchi and A. Benjebbour, "Non-orthogonal multiple access (NOMA) with successive interference cancellation," *IEICE Trans. Commun.*, vol. E98-B, no. 3, pp. 403–414, Mar. 2015.

42. Z. Ding, Z. Yang, P. Fan, and H.V. Poor, "On the performance of non-orthogonal multiple access in 5G systems with randomly deployed users," *IEEE Signal Process. Lett.*, vol. 21, no. 12, pp. 1501–1505, Dec. 2014.

43. X. Chen, A. Benjebbour, Y. Lan, A. Li, and H. Jiang, "Impact of rank optimization on downlink non- orthogonal multiple access (NOMA) with SU-MIMO," in *Proc. IEEE Int. Conf. Commun. Syst. (ICCS)*, Nov. 2014, pp. 233–237.

44. C.H. Liu, and D.C. Liang, "Heterogeneous networks with power-domain NOMA: Coverage, throughput, and power allocation analysis," *IEEE Trans. Wireless Commun.*, vol. 17, no. 5, pp. 3524–3539, 2018.

45. A.A. Nasir, H.D. Tuan, T.Q. Duong, and M. Debbah, "NOMA throughput and energy efficiency in energy harvesting enabled networks," *IEEE Trans. Commun.*, vol. 67, no. 9, pp. 6499–6511, 2019.

46. A. Pastore, and M. Navarro, "A fairness–throughput tradeoff perspective on NOMA multiresolution broadcasting," *IEEE Trans. Broadcasting*, vol. 65, no. 1, pp. 179–186, 2018.

47. H. Xing, Y. Liu, A. Z. Ding, and H.V. Poor, "Optimal throughput fairness tradeoffs for downlink non-orthogonal multiple access over fading channels," *IEEE Trans. Wireless Commun.*, vol. 17, no. 6, pp. 3556–3571, 2018.

48. X. Liu, Y. Wang, S. Liu, and J. Meng, "Spectrum resource optimization for NOMA-based cognitive radio in 5G communications," *IEEE Access*, vol. 6, pp. 24904–24911, 2018.

49. R. L. G. Cavalcante, S. Stanczak, M. Schubert, A. Eisenlatter, and U. Turke, "Toward energy-efficienct 5G wireless communications technologies," *IEEE Signal Process. Mag.*, vol. 13, no. 11, pp. 24–34, 2014.

50. H. Zhang, J. Zhang, and K. Long, "Energy efficiency optimization for NOMA UAV network with imperfect CSI," *IEEE J. Sel. Areas Commun.*, vol. 38, no. 12, pp. 2798–2809, 2020.

51. M.W. Baidas, E. Alsusa, and Y. Shi, "Resource allocation for SWIPT-enabled energy-harvesting downlink/uplink clustered NOMA networks," *Computer Networks*, vol. 182, 2020.

52. Y. Zuo, X. Zhu, Y, Jiang, Z. Wei, H. Zeng, and T. Wang, Energy efficiency and spectral efficiency tradeoff for multicarrier NOMA systems with user fairness. In 2018 IEEE/CIC International Conference on Comunications in China (ICCC), pp. 666–670, 2018.

53. M.W. Baidas, "Distributed energy-efficiency maximization in energy-harvesting uplink NOMA relay ad-hoc networks: Game-theoretic modeling and analysis," *Physical Communication*, vol. 43, 2020.

54. Q.H. Tran, C.V. Phan, and Q.T. Vien, "Performance analysis of power-splitting relaying protocol in SWIPT based cooperative NOMA systems," *Journal on Wireless Communications and Networking*, 2021, 110. DOI: *10.21203/rs.3.rs-43624/v2*
55. H.Q. Tran, T.T. Nguyen, C. V. Phan, Q.T. Vien, "Power-splitting relaying protocol for wireless energy harvesting and information processing in NOMA systems," *IET Communications*, vol. 13, no. 14, pp. 2132–2140, 2019.
56. Joydev Ghosh, In-Ho Ra, Saurabh Singh, Huseyin Haci, Khaled AlUtaibi, and Sadiq M. Sait, On the comparison of optimal NOMA and OMA in a paradigm shift of emerging technologies. *TechRxiv*. 2022. DOI: *10.36227/techrxiv.18972947.v1*
57. M. Al-Imari, M. A. Imran, and R. Tafazolli, "Low density spreading for next generation multicarrier cellular systems," in *Proc. IEEE Int. Conf. Future Commun. Networks (ICFCN)*, Apr. 2012, pp. 52–5.
58. S. Choudhury, and J.D. Gibson, "Information transmission over fading channels," in *Proc. IEEE Global Telecommun. Conf. (GLOBECOM)*, Nov. 2007, pp. 3316–3321.
59. S.M.R. Islam and K.S. Kwak, "Outage capacity and source distortion analysis for NOMA users in 5G systems," *Electronics Letters*, vol. 52, no. 15, pp. 1344–1345, Jul. 2016.
60. B. Lee, D. Park, and H. Seo, *Wireless Communications Resource Management*. Wiley, 2009.
61. L. Lei, D. Yuan, C. K. Ho, and S. Sun, "Joint optimization of power and channel allocation with non- orthogonal multiple access for 5G cellular systems," in *Proc. IEEE Global Telecommun. Conf. (GLOBECOM)*, Dec. 2015, pp. 1–6.
62. M. Zeng, G. I. Tsiropoulos, O. A. Dobre, and M. H. Ahmed, "Power allocation for cognitive radio networks employing non-orthogonal multiple access," in Proc. IEEE Global Telecommun. Conf. (GLOBECOM), Dec. 2016, In Press.
63. B. Kim, S. Lim, H. Kim, S. Suh, J. Kwun, S. Choi, C. Lee, S. Lee, and D. Hong, "Non-orthogonal multiple access in a downlink multiuser beamforming system," in *Proc. IEEE Military Commun. Conf.*, San Diego, CA, USA, Nov. 2013.
64. R. C. Kizilirmak, C. R. Rowell, and M. Uysal, "Non-orthogonal multiple access (NOMA) for indoor visible light communications," in *Proc. 4th Int. Workshop Optical Wireless Commun. (IWOW)*, Sep. 2015, pp. 98–101.
65. M. J. Hagh and M. R. Soleymani, "Raptor coding for non-orthogonal multiple access channels," in *Proc. IEEE Int. Conf. Commun. (ICC)*, Jun. 2011, pp. 1–6.
66. S. Park and D. Cho, "Random linear network coding based on non orthogonal multiple access in wireless networks," *IEEE Commun. Lett.*, vol. 19, no. 7, pp. 1273–1276, Jul. 2015.
67. S. Rangan, T. S. Rappaport, and E. Erkip, "Millimeter-wave cellular wireless networks: potentials and challenges," *Proc. IEEE*, vol. 102, no. 3, pp. 366–385, Mar. 2014.
68. Z. Ding, P. Fan, and H. V. Poor, "Impact of user pairing on 5G non-orthogonal multiple access downlink transmissions," *IEEE Trans. on Veh. Technol.*, In Press, Sep. 2015.
69. F. Liu, P. Mähönen, and M. Petrova, "Proportional fairness-based user pairing and power allocation for non- orthogonal multiple access," in *Proc. IEEE 26th Annu. Int. Symp. Personal, Indoor, and Mobile Radio Commun. (PIMRC)*, Sep. 2015, pp. 1127–1131.
70. T. Seyama, T. Dateki, H. Seki, "Efficient selection of user sets for downlink non-orthogonal multiple access," in *Proc. IEEE 26th Annu. Int. Symp. Personal, Indoor, and Mobile Radio Commun. (PIMRC)*, Sep. 2015, pp. 1062–1066.
71. T. S. Rappaport, S. Shu, R. Mayzus, Z. Hang, Y. Azar, K. Wang, G.N. Wong, J.K. Schulz, M. Samimi, and F. Gutierrez, "Millimeter wave mobile communications for 5G cellular: It will work!," *IEEE Access*, vol. 1, pp. 335–349, 2013.
72. R. Q. Hu and Y. Qian, *Heterogeneous Cellular Networks*. Wiley, 2013

73. R. Q. Hu and Y. Qian, "An energy efficient and spectrum efficient wireless hetero-geneous network framework for 5G systems," *IEEE Commun. Mag.*, vol. 52, no. 5, pp. 94–101, May. 2014.

74. Z. Ding, F. Adachi, H. V. Poor, "The application of MIMO to non-orthogonal multiple access," *IEEE Trans. Wireless Commun.*, vol. 15, no. 1, pp. 537–552, Jan. 2016.

75. J. Kim, J. Koh, J. Kang, K. Lee, and J. Kang, "Design of user clustering and precod-ing for downlink non- orthogonal multiple access (NOMA)," in *Proc. IEEE Military Commun. Conf. (MILCOM)*, Oct. 2015, pp. 1170–1175.

76. G. Yuan, X. Zhang, W. Wang, and Y. Yang, "Carrier aggregation for LTE-advanced mobile communication systems," *IEEE Commun. Mag.*, vol. 48, no. 2, pp. 88–93, Feb. 2010.

77. S. Liu and C. Zhang, "Downlink non-orthogonal multiple access system with limited feedback channel," in *Proc. IEEE 7th Int. Conf. Wireless Commun. Signal Process. (WCSP)*, Nanjing, Oct. 2015, pp. 1–5.

78. J. B. Kim and I. H. Lee, "Non-orthogonal multiple access in coordinated direct and relay transmission," *IEEE Commun. Lett.*, vol. 19, no. 11, pp. 2037–2040, Nov. 2015.

79. V. Kalokidou, O. Johnson, and R. Piechocki, "A hybrid TIM-NOMA scheme for the SISO Broadcast Channel," in *Proc. IEEE Int. Conf. Commun. Workshop (ICCW)*, London, Jun. 2015, pp. 387–392.

80. X. Xiong, W. Xiang, K. Zheng, H. Shen and X. Wei, "An open source SDR-based NOMA system for 5G networks," *IEEE Wireless Commun.*, vol. 22, no. 6, pp. 24–32, Dec. 2015.

81. K. Higuchi, and A. Benjebbour, "Non-orthogonal multiple access (NOMA) with suc-cessive interference cancellation for future radio access," *IEEE Trans. Commun.*, vol. 98, no. 3, pp. 403–414, 2015.

82. Q. Sun, S. Han, I. Chin-Lin, and Z. Pan, On the ergodic capacity of MIMO NOMA systems. *IEEE Wireless Commun. Lett.*, vol. 4, no. 4, pp. 405–408, 2015.

83. Y. Saito, A. Benjebbour, Y. Kishiyama, and T. Nakamura, "System-level performance evaluation of downlink non-orthogonal multiple access (NOMA)," in *Proc. IEEE 24th Annu. Inter. Sympos. Personal, Indoor Mobile Radio Commun. (PIMRC)*, London, UK, 8–11 September 2013; pp. 611–615.

84. Y. Saito, A. Benjebbour, Y. Kishiyama, and T. Nakamura, "System-level performance of downlink non-orthogonal multiple access (NOMA) under various environments," in *Proc. IEEE 77th Veh. Technol. Conf. (VTC Spring)*, Glasgow, UK, 11–14 May 2015; pp. 1–5.

85. A. Benjebbovu, A. Li, Y. Saito, Y. Kishiyama, A. Harada, and T. Nakamura, "System-level performance of downlink NOMA for future LTE enhancements," in *Proceedings of the 2013 IEEE Globecom Workshops (GC Wkshps)*, Atlanta, GA, USA, 9–13 December 2013; pp. 66–70.

86. N. Zhao, W. Lu, M. Sheng, Y. Chen, J. Tang, F.R. Yu, and K.K. Wong, "UAV-assisted emergency networks in disasters," *IEEE Wireless Commun.*, vol. 26, no. 1, pp. 45–51, 2019.

87. E.W. Frew, and T.X. Brown, "Airborne communication networks for small unmanned aircraft systems," *Proceedings of the IEEE*, vol. 96, no. 12, 2008.

88. Z. Ding, Y. Liu, J. Choi, Q. Sun, M. Elkashlan, and H.V. Poor, Application of non-orthogonal multiple access in LTE and 5G networks. arXiv preprint arXiv:1511.08610, 2015.

89. W.K. New, C.Y. Leow, K.N.Y. Sun, and Z. Ding, "Application of NOMA for cellular-connected UAVs: opportunities and challenges," *SCIENCE CHINA Information Sciences*, March 2021. DOI: *10.1007/s11432-020-2986-8*

90. Y. Liu, G. Pan, H. Zhang, and M. Song, "On the capacity comparison between MIMO-NOMA and MIMO- OMA," *IEEE Access*, vol. 4, pp. 2123–2129, Sept. 2016.

91. P. Xu, Z. Ding, X. Dai, and H. V. Poor, "A new evaluation criterion for non-orthogonal multiple access in 5G software defined networks," *IEEE Access*, vol. 3, pp. 1633–1639, Sept. 2015.
92. Z. Chen, Z. Ding, P. Xu, and X. Dai, "Optimal precoding for a QoS optimization problem in two-user MISO- NOMA downlink," *IEEE Commun. Lett.*, vol. 20, no. 6, pp. 1263–1266, Jun. 2016.
93. T. S. Rappaport, S. Sun, R. Mayzus, H. Zhao, Y. Azar, K. Wang, G. N. Wong, J. K. Schulz, M. Samimi, and F. Gutierrez, "Millimeter wave mobile communications for 5G cellular: It will work!" *IEEE Access*, vol. 1, pp. 335–349, 2013.
94. Enable higher frequency spectrum for future wireless. Federal Communications Commission Report (FCC 16-89), Jul. 2016.
95. E. Nussbaum, "Integrated broadband networks and services of the future," in *Proc. Optical Fiber Commun. Conf.*, Reno, Nevada, US, Jan. 1987.
96. B. Wang, L. Dai, Z. Wang, N. Ge, and S. Zhou, "Spectrum and energy-efficient beam-space MIMO-NOMA for millimeter-wave communications using lens antenna array," *IEEE J. Sel. Areas Commun.*, vol. 35, no. 10, pp. 2370–2382, 2017.
97. Z. Zhang, Z. Ma, M. Xiao, Z. Ding, P. Fan, "Full-duplex device-to-device-aided cooperative non-orthogonal multiple access," *IEEE Trans. Veh. Technol.*, vol. 66, pp. 4467–4471, 2016.
98. L. Pei, Z. Yang, C. Pan, W. Huang, M. Chen, M. Elkashlan, and A. Nallanathan, "Energy-efficient D2D communications underlaying NOMA-based networks with energy harvesting," *IEEE Commun. Lett.*, vol. 22, pp. 914–917, 2018.
99. Y.B. Song, H.S. Kang, and D.K. Kim, "5G cellular systems with D2D assisted NOMA relay," In *Proceedings of the URSI Asia-Pacific Radio Science Conference (URSI AP-RASC)*, Seoul, Korea, 21–25 August 2016, pp. 1–3.
100. A. Anwar, B.C. Seet, and Z. Ding, "Non-orthogonal multiple access for ubiquitous wireless sensor networks," vol. 18, pp. 516, 2018.
101. D.T. Do and C.B. Le, "Application of NOMA in wireless system with wireless power transfer scheme: Outage and ergodic capacity performance analysis," *Sensors*, vol. 18, pp. 3501, 2018.
102. H. Nguyen, T. Nguyen, P.T. Tin, and M. Voznak, "Outage performance of time switching energy harvesting wireless sensor networks deploying NOMA," in *Proceedings of the IEEE 20th International Conference on e-Health Networking, Applications and Services (Healthcom)*, Ostrava, Czech Republic, 17–20 September 2018; pp. 1–4.
103. Z. Hu, L. Xu, L. Cao, S. Liu, Z. Luo, J. Wang, X. Li, and L. Wang, "Application of non-orthogonal multiple access in wireless sensor networks for smart agriculture," *IEEE Access*, vol. 7, pp. 87582–87592, 2019.

6 Multiple Access Schemes for 5G and Beyond Wireless Networks

Dileep Kumar, Niraj Pratap Singh, and Gaurav Verma

6.1 INTRODUCTION

5G wireless networks have become increasingly important in our everyday lives, facilitating unhindered connectivity and worldwide communication. These networks have come into existence due to the widespread use of mobile stations and the increasing desire for excellent multimedia services, providing fast data rates, reduced latency, and better network efficiency. The evolution of wireless communication networks with key technology is shown in Figure 6.1. It is important for wireless networks, especially those in 5G and beyond, to develop efficient and reliable multiple access schemes, which are responsible for sharing limited wireless resources among multiple users. Multiple access schemes for 5G/6G wireless networks are covered in this chapter, including traditional approaches, as well as recent advancements and future prospects.

6.1.1 TRADITIONAL MULTIPLE ACCESS SCHEMES

The word multiple access (MA) refers to a technique that facilitates simultaneous access to the wireless network by multiple users without interfering with each other. The milestone of MA schemes is shown in Figure 6.2. In traditional wireless networks, several famous MA schemes have been used, including frequency-division multiple access (FDMA), "time-division multiple access (TDMA)," and "code-division multiple access (CDMA)." They also have limitations in terms of synchronization requirements, latency, spectral efficiency, flexibility, and complexity, which may impact their performance in certain scenarios. Advanced MA schemes have been developed to overcome some of these limitations and provide improved performance in 5G/6G wireless networks.

6.1.2 ADVANCED MULTIPLE ACCESS SCHEMES FOR 5G AND BEYOND

To meet the rising demand for larger data rates, less latency, and improved network efficiency in 5G and beyond wireless networks, several advanced MA

DOI: 10.1201/9781003407836-6

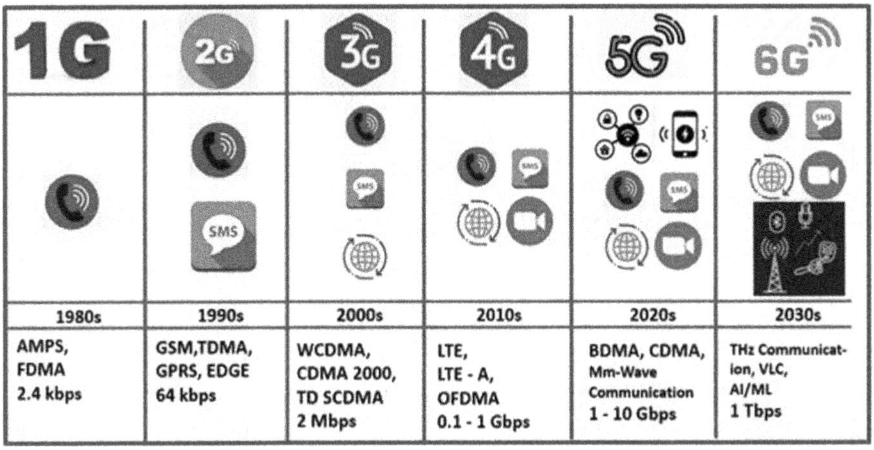

1G	2G	3G	4G	5G	6G
1980s	1990s	2000s	2010s	2020s	2030s
AMPS, FDMA 2.4 kbps	GSM,TDMA, GPRS, EDGE 64 kbps	WCDMA, CDMA 2000, TD SCDMA 2 Mbps	LTE, LTE - A, OFDMA 0.1 - 1 Gbps	BDMA, CDMA, Mm-Wave Communication 1 - 10 Gbps	THz Communication, VLC, AI/ML 1 Tbps

FIGURE 6.1 The evolution of wireless communication networks.

schemes have been proposed and researched. Some popular MA schemes such as "OFDMA," "NOMA," "SDMA," "PDMA," "RSMA," "RMA," and "D-OMA" are discussed in this chapter.

In this chapter, the above MA schemes including their design principles, performance analysis, and implementation challenges are discussed one by one.

6.2 SPACE DIVISION MULTIPLE ACCESS (SDMA)

SDMA is a method of communication that facilitates the sharing of a single frequency band and time slot among multiple users. This is achieved by dividing the available physical spaces or spatial channels among the different users. Figure 6.3 demonstrates how spot beam antennas are used by SDMA to accommodate various users [1].

SDMA aims to enhance wireless communication by utilizing multiple antennas at both ends of the transmission to create distinct spatial channels for each user. Each user is nominated a different spatial channel, and the signals transmitted by each user are spatially separated so that they do not interfere with each other.

This enables multiple individuals to transmit messages concurrently without interference, thus boosting the communication system's capacity. The spatial channels can be created using various techniques such as beamforming, which directs the transmission power in specific directions, or spatial filtering, which separates signals based on their arrival angles.

SDMA is particularly useful in environments with a high density of users, such as urban areas or stadiums, where traditional communication methods like TDMA or FDMA may not be able to provide sufficient capacity.

SDMA is also used in WLANs and cellular systems to improve system capacity and user experience.

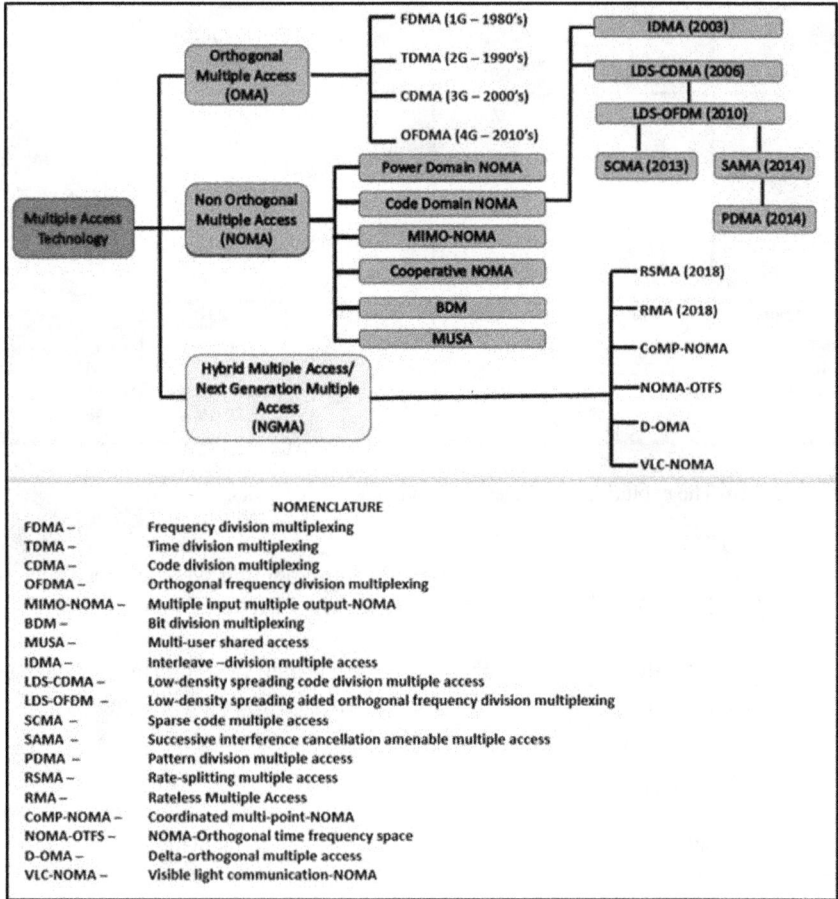

FIGURE 6.2 The milestone of multiple access schemes.

6.3 ORTHOGONAL FREQUENCY DIVISION MULTIPLE ACCESS (OFDMA)

OFDMA is an advancement on OFDM. In contrast to OFDM, which is single-user, OFDMA is multi-user, as seen in Figure 6.4. For small data packets or several endpoints, it provides a three times greater throughput than single-user OFDM [2]. Multiple endpoints are simultaneously sent frames by combining transmissions and using OFDMA. Transmission with a shorter latency improves efficiency. For applications involving Internet of Things (IoT) devices, video, online gaming, and automation, OFDMA is the best choice. A popular method for accessing 4G and 5G wireless networks is dividing the available frequency band into smaller subcarriers, which are then allocated to individual users for communication. OFDMA technology enables several users to communicate at

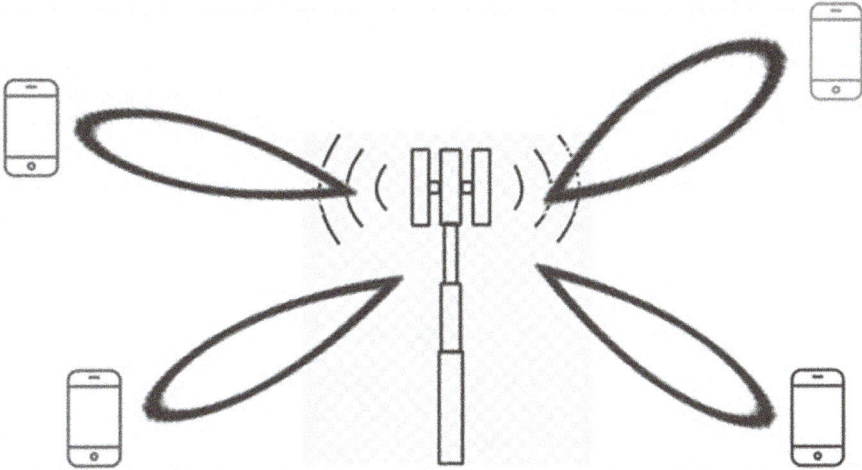

FIGURE 6.3   A spot beam-based spatial base station antenna that serves many users.

the same frequency band, utilizing spectrum more efficiently thanks to subcarrier orthogonality. OFDMA provides high SE, protectiveness against frequency-selective fading, and changeability in resource allocation, making it suitable for diverse communication scenarios, such as cellular networks, Wi-Fi, and broadband access.

Future systems should take into account some of the drawbacks of OFDM, such as sensitivity to timing and frequency offsets, cyclic prefix (CP) overhead, high peak-to-average power ratio (PAPR), and high out-of-band (OOB) emissions.

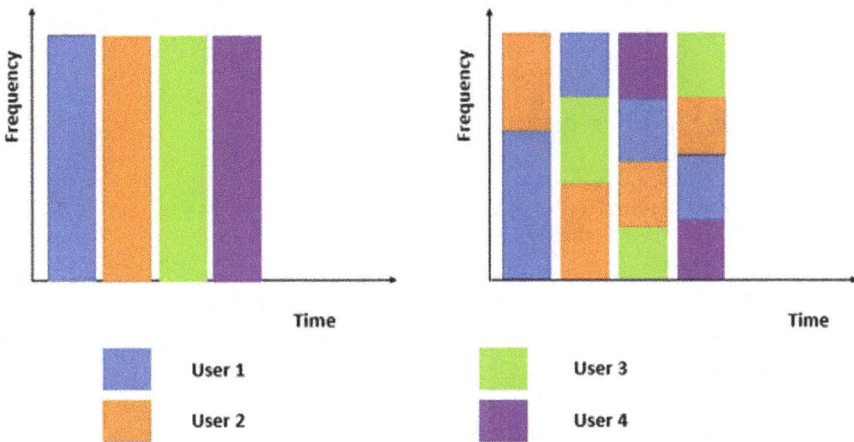

FIGURE 6.4   OFDM vs. OFDMA.

6.4   NON-ORTHOGONAL MULTIPLE ACCESS (NOMA)

NOMA is a relatively new technology in wireless communications that has attracted a lot of attention in recent years. A NOMA network allows multiple users to access the same frequency and time resources at the same time. Especially in scenarios with a high number of users and a limited amount of spectrum resources, NOMA has the spark to increase spectral efficiency, capacity, and energy efficiency [3–6]. NOMA can be classified as shown in Figure 6.2. The basic concept of power domain NOMA is explained in the next sections.

6.4.1   Basic Concept of Downlink NOMA

It is the idea behind downlink NOMA to allocate dissimilar power levels according to the conditions of the channel to different users. Clients/users having weak channel conditions receive high power levels, and clients/users having strong channel conditions are assigned low power levels. Base stations can transmit data to multiple users simultaneously while maintaining a high quality of service for each.

An illustration of the conventional NOMA Downlink system can be seen in Figure 6.5.

The system has one base station (BS) and K users. kth user (k $\in$ {1,2,3......K}) is assumed to be located at the kth distance position from the BS for simplicity. The BS transmits a signal for the kth user the transmitted power Ps. Then, the transmitted signal over W bandwidth, S, can be written as

$$S = \sum_{i=1}^{K} \sqrt{a_i P_s}\, x_i,$$ (6.1)

where a_i is the power coefficient allocated for user I and $a_1 \geq a_2 \geq$ .... a_k.

Without any loss of generality, the channel gains can be assumed to be in ascending order as $|h_1|^2 \leq |h_2|^2 \leq$ .... $|h_K|^2$. The signal received at the kth user can be expressed as

$$y_k = h_k S + n_k = h_k \left(\sum_{i=1}^{K} \sqrt{a_i P_s}\, x_i \right) + n_k,$$ (6.2)

where n_k is "zero mean complex additive Gaussian noise" with a variance of σ^2; that is $n_k \sim CN(0, \sigma^2)$.

6.4.1.1   Signal to Interference and Noise Ratio (SINR) Analysis

It is challenging to do successive interference cancellation (SIC) for the client/user with lesser channel gain (whose distance from the BS is greater). This is due to the received signal power which is lower. In this instance, the intended signal is detected directly and at a greater strength without the need of SIC. It should be noted that after conducting SIC in the receiver of the kth user, inter user interference still exists and is not eliminated by SIC.

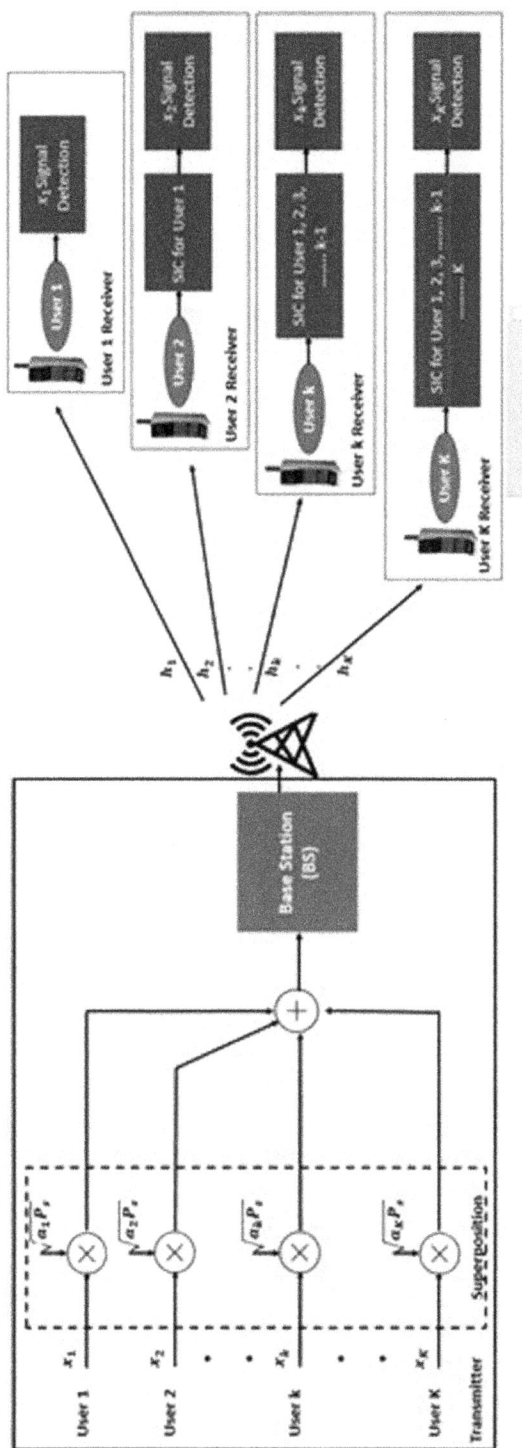

FIGURE 6.5   Conventional downlink NOMA system.

As a result, after obtaining the signal y_k at the kth user, SIC successively detects and extracts from y_k all the signals from x_{k+1} to x_K. The obtained SINR of the kth user in this instance may be represented as

$$SINR_k = \frac{a_k P_s |h_k|^2}{P_s |h_k|^2 \sum_{i=k+1}^{K} a_i + \sigma^2}$$

$$SINR_k = \frac{a_k P |h_k|^2}{P |h_k|^2 \sum_{i=k+1}^{K} a_i + 1} \tag{6.3}$$

where $P = \frac{P_s}{\sigma^2}$ denotes the SNR.

Then SINR of Kth user is expressed as

$$SINR_K = a_K P |h_K|^2 . \tag{6.4}$$

6.4.1.2 Sum Rate Analysis

The possible data rate (throughput) of the kth user for downlink NOMA can be expressed as

$$R_k^{NOMA-d} = log_2 (1 + SINR_k)$$

$$R_k^{NOMA-d} = log_2 \left(1 + \frac{a_k P |h_k|^2}{P |h_k|^2 \sum_{i=k+1}^{K} a_i + 1} \right) . \tag{6.5}$$

Therefore "sum-rate of downlink NOMA" may be written as

$$R_{SUM}^{NOMA-d} = \sum_{k=1}^{K} log_2 (1 + SINR_k)$$

$$R_{SUM}^{NOMA-d} = \sum_{k=1}^{K} log_2 \left(1 + \frac{a_k P |h_k|^2}{P |h_k|^2 \sum_{i=k+1}^{K} a_i + 1} \right)$$

$$R_{SUM}^{NOMA-d} = \sum_{k=1}^{K-1} log_2 \left(1 + \frac{a_k P |h_k|^2}{P |h_k|^2 \sum_{i=k+1}^{K} a_i + 1} \right) + log_2 \left(1 + a_K P |h_K|^2 \right) \tag{6.6}$$

$$R_{SUM}^{NOMA-d} = \sum_{k=1}^{K-1} log_2 \left(1 + \frac{a_k}{\sum_{i=k+1}^{K} a_i + \frac{1}{P |h_k|^2}} \right) + log_2 \left(1 + a_K P |h_K|^2 \right)$$

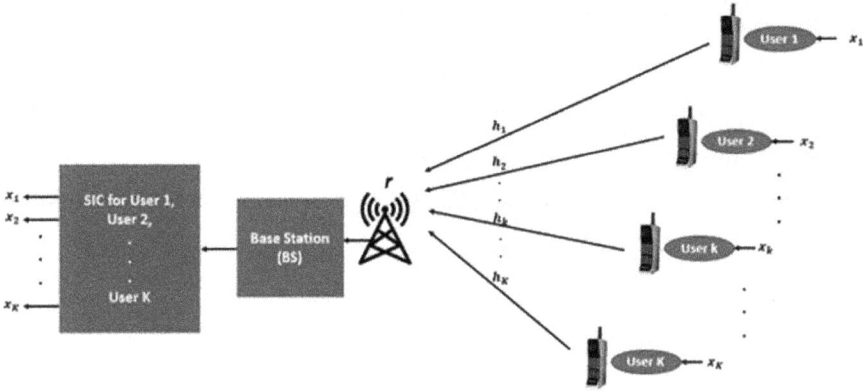

FIGURE 6.6 Conventional uplink NOMA system.

At high SNR, that is P → ∞, then

$$R_{SUM}^{NOMA-d} \approx log_2\left(P|h_K|^2\right). \tag{6.7}$$

6.4.2 BASIC CONCEPT OF UPLINK NOMA

In an uplink NOMA network, mobile clients/users send their signals to the BS. Refer to Figure 6.6 for a visual representation. Iterations are done at the BS SIC to find the signal of mobile users. The BS sends out power allocation coefficients to mobile clients/users, assuming that their downlink and uplink channels do not overlap. For synchronous uplink NOMA, the obtained signal at the BS may be represented as

$$r = \sum_{i=1}^{K} h_i \sqrt{a_i P_{max}}\, x_i + n, \tag{6.8}$$

where P_{max} is the "maximum transmitted power" assumed to be same for all clients/users, hi is the "channel coefficient" of the ith user, and n is "zero mean complex additive Gaussian noise with variance" of σ^2; that is, n ~ CN $(0, \sigma^2)$.

6.4.2.1 SINR Analysis

The BS decodes the users' signal in the order of decreasing power coefficients, from which the SINR for the kth client/user (where k ≠ 1) is then determined, excepting the first user as

$$SINR_k = \frac{a_k P|h_k|^2}{P\displaystyle\sum_{i=1}^{k-1} a_i|h_i|^2 + 1}, \tag{6.9}$$

where $P = \frac{P_{max}}{\sigma^2}$. Now, the first user SINR is expressed as

$$SINR_1 = a_1 P |h_1|^2. \tag{6.10}$$

6.4.2.2 Sum Rate Analysis

The achievable data rate (throughput) of uplink NOMA for kth user may be expressed as

$$R_k^{NOMA-U} = log_2\left(1 + SINR_k\right)$$

$$R_k^{NOMA-U} = log_2\left(1 + \frac{a_k P |h_k|^2}{P \sum_{i=1}^{k-1} a_i |h_i|^2 + 1}\right). \tag{6.11}$$

Therefore "sum-rate of uplink NOMA" may be written as

$$R_{SUM}^{NOMA-U} = \sum_{k=1}^{K} log_2\left(1 + SINR_k\right)$$

$$R_{SUM}^{NOMA-U} = log_2\left(1 + a_1 P |h_1|^2\right) + \sum_{k=2}^{K} log_2\left(1 + \frac{a_k P |h_k|^2}{P \sum_{i=1}^{k-1} a_i |h_i|^2 + 1}\right). \tag{6.12}$$

$$R_{SUM}^{NOMA-U} = log_2\left(1 + P \sum_{k=1}^{K} a_k |h_k|^2\right)$$

At high SNR, that is $P \to \infty$, then

$$R_{SUM}^{NOMA-U} \approx log_2\left(P \sum_{k=1}^{K} |h_k|^2\right). \tag{6.13}$$

6.4.3 IMPERFECTNESS IN NOMA

Ideal cancellation in the receiver with SIC has been discussed in our previous parts. In reality, precision in the subtraction of the decoded signal from the received signal presents a significant challenge. An idea in NOMA with cancellation error in the SIC receiver will be review in this part.

Here, the downlink is simply taken into account, but the talks might easily be expanded to include the uplink. It is notable that the SIC receiver performs an iterative decoding process when attempting to acquire a specific information signal. In SIC, it is necessary to re-create the initial individual waveform following the decoding of the signal to eliminate it from the received signal. In practice, errors leading to cancellations are likely to occur even though the process can theoretically be carried out flawlessly.

In downlink, the "SNR for the kth user with cancellation error" is written as [7]

$$SINR_k = \frac{a_k P |h_k|^2}{P|h_k|^2 \left(\sum_{i=k+1}^{K} a_i + \epsilon \sum_{i=1}^{k-1} a_i + 1 \right)},$$

(6.14)

where the remaining part of the signal from the cancelled message is represented by the cancellation error ϵ. Since complete cancellation is assumed in the preceding section, the error term in the denominator is left out.

6.4.4 SPECTRAL EFFICIENCY AND ENERGY EFFICIENCY

The network's throughput performance was taken into account in the majority of analyses thus far. In this part, the energy efficiency (EE) of NOMA systems is studied in addition to the spectral efficiency (SE) of NOMA systems. Both the power used for the information waveform and the network's static power usage as a result of the power amplifiers are included in this study.

The sum of the power used by the circuits and the data signal power can be used to represent the total power used at the transmitter (mostly by power amplifiers). The total power used by the BS can then be calculated by taking into account the downlink as [8]

$$P_{total} = P_T + P_{static},$$

(6.15)

where P_T is the "total signal power as mentioned earlier" and Pstatic is the "power consumed by the circuitry."

EE is defined as "the sum rate over the total consumed power of the base station" [6]

$$EE = \frac{R_T}{P_{total}} = SE \frac{W}{P_{total}} \left(bits/joule \right),$$

(6.16)

where SE represents spectral efficiency $\left({R_T}/{W} \right)$ in bits per second (bps/Hz).

The extensive research conducted in the PD-NOMA over the past decade has been remarkable. At present some recent work is being done in the field of PD-NOMA.

6.4.5 OTHER NOMA SCHEMES

The basic concept of **Power domain NOMA** is explained in the previous section. The huge research has been carried out in the NOMA in the last decade. Some popular NOMA schemes are given below except power domain NOMA.

6.4.5.1 Code Domain Non-Orthogonal Multiple Access (CD-NOMA)

CD-NOMA is a promising technology for improving the spectral efficiency and capacity of wireless communication networks. Various technological advancements in code domain NOMA have been carried out in past few years. Some popular CD-NOMA are explained below:

Sparse code multiple access (SCMA) is a type of CD-NOMA technique that enables many clients/users to utilise the same time and frequency resources through sparse spreading sequences. Recent research conducted by the author of [3] evaluates the efficiency of the low-rate turbo-coded power domain (LR-TC-PD)-SCMA NOMA system in relation to various recognized multi-dimensional constellation (MDC) designs. To ensure the distinctiveness of users' symbols in overloaded systems, the authors suggest a particular design for the rotation of users, which is tailored to each user, using quaternion rotation matrices [3]. Recent advancements in SCMA include the use of deep learning (DL) algorithms to optimize the codebook design and user scheduling for improved system performance. DLT has been implemented in the SCMA detector to increase the effectiveness of detection. To reduce the load on the neural network, [4] has presented a knowledge-based DL detection strategy (K-DLD) that includes prior information into SCMA detection [4]. Multi-carrier NOMA (MC-NOMA), a code domain NOMA technique, allows many clients/users to access the same frequency resources using OFDM. Recent advancements in MC-NOMA have seen the emergence of sophisticated signal processing methods such as interference cancellation and collaborative resource allocation to boost system performance. Using the "multi-carrier cell-less nonorthogonal multiple access (MC-CL-NOMA)" architecture, a framework has been suggested that increases the overall throughput of a Next-Generation Wireless Network (NGWN) [5]. The framework carries out this action by feeding an ensemble metaheuristic algorithm with scenario data [5]. For uplink MC-NOMA-based "two-tier heterogeneous networks (HetNets)" with wireless resource allocation have been studied [6]. By allocating different MC-NOMA users (NUs) different subchannels while enabling numerous NUs to use the common subchannel, the author has taken advantage of channel diversity. To maximise EE under the limitations of a lowest quality of service (QoS) requirement, the maximum permitted transmit power for each NU, and consecutive interference cancellation, the author formulates a joint NU clustering, power, and subchannel allocation problem [6].

Interleave-division multiple access (IDMA) is a wireless communication technology (WCT) that enables many clients/users to utilise the same frequency band by interspersing data packets of various clients/users. In IDMA, the interleaving process separates the data packets of different clients/users into smaller sub-packets and then interleaves them. This results in many clients/users transmitting concurrently over a similar frequency band, without causing interference to each other. IDMA is different from other MA technologies such as TDMA and CDMA, which rely on dividing the frequency band into time slots or assigning different codes to particular clients/users. IDMA is particularly suitable for low power and low data rate communication systems, such as machine-to-machine (M2M) communication and IoT. The amalgamated system, which incorporates all the benefits of intelligent reflecting surfaces (IRS), IDMA, and NOMA, has been proposed by [9]. It is known as an IRS-aided full duplex IDMA communication system. IDMA has several advantages over other multiple access technologies, including lower power consumption, improved SE, and better resistance to interference. However, IDMA also presents several challenges, including the need for sophisticated signal processing algorithms to decode the interleaved sub-packets and the potential for increased latency due to the interleaving process.

Low-density spreading code division multiple access (LDS-CDMA) is a type of CDMA technology used in wireless communication system (WCS). It is built upon the idea of distributing the signal across a wide frequency range by utilizing distinct codes for each user, thereby permitting many clients/users to share the common frequency channel without causing any interference. In LDS-CDMA, a low-density parity-check (LDPC) code is used for spreading the signal, which has a low density of ones in the code matrix. The LDPC code is a type of error-correcting code that is used to improve the reliability of communication by detecting and correcting errors in the transmitted data. [10] studies an algebraic design-based code-domain NOMA approach. A projective geometry-based approach to design an improved low-density spreading (LDS) sequence has been proposed by the author. The author's proposed augmented LDS code set demonstrates improved BER results compared to the current LDS designs in both frequency-nonselective Rayleigh fading (FNSRF) and additive white gaussian noise (AWGN) channels. Additionally, the author showed how the shortest distance is necessary to achieve the best bit error rate (BER) [10]. One of the key benefits of LDS-CDMA in contrast with other CDMA technologies is its capacity to hold a large number of clients/users in a single cell, without creating interference or diminishing the QoS. This is accomplished by utilizing a distinct spreading code for each user, enabling the receiver to identify and segregate the signals from different clients/users. LDS-CDMA is used worldwide in the newest wireless communication systems, such as wireless sensor networks, satellite communication systems, and 4G and 5G cellular networks. It is also used in other applications, such as wireless LANs and RFID systems.

Low-density spreading aided OFDMA (LD-OFDMA) is a wireless communication technique that mix the advantages of OFDM and LDS to increase spectral efficiency and reduce interference. In traditional OFDM, the existing bandwidth is divided into multiple subcarriers that are orthogonal to each other, which allows many clients/users to transmit data concurrently devoid of interfering with each other. However, in high-mobility scenarios, the frequency-selective fading can cause the subcarriers to experience different channel conditions, leading to inter-carrier interference (ICI) and a reduction in performance. LDS, on the other hand, is a technique that uses sparse spreading sequences to reduce the number of active subcarriers in the OFDM signal, thereby increasing the SE and reducing interference. The sparse spreading sequences are selected in such a manner that the subcarriers having poor channel conditions are deactivated, while the subcarriers with good channel conditions are kept active. LD-OFDMA combines these two techniques by applying LDS to the subcarriers in OFDM. This reduces the number of active subcarriers and increases the span between them, which reduces ICI and improves the system's performance in high-mobility scenarios. Furthermore, LD-OFDMA allows for a more flexible allocation of subcarriers to clients/users, which improves the system's capacity and SE. Overall, LD-OFDMA is a highly promising approach for coming WCS that demand high spectral efficiency and robustness in scenarios with high mobility.

Successive interference cancellation amenable multiple access (SAMA): SIC is a technique used in MA schemes to overcome interference and increase the capacity of wireless communication systems. It is commonly employed in "CDMA" and "OFDMA" systems. While SIC is a general concept applicable to various multiple access schemes. AMA is a recent MA scheme that incorporate the advantages of CDMA and SIC. It aims to gain the capacity and efficiency of WCN by exploiting interference cancellation techniques. The AMA uses a combination of randomization and successive interference cancellation to enable multiple clients/users to transmit simultaneously. In AMA, every user is entitled to a distinctive identifier sequence or code, much like CDMA. The clients/users send their signals simultaneously, leading to interference at the receiver. However, with the aid of SIC, the receiver can decode the signals of multiple clients/users by iteratively canceling out the interference from already decoded signals. The SIC process in AMA involves multiple stages. In each stage, the receiver selects one user's signal with the highest received power and decodes it by treating other signals as interference. Once decoded, the receiver subtracts the estimated signal from the obtained signal, effectively canceling its contribution from the interference set. The process continues iteratively, decoding one user at a time until all clients/users' signals are recovered or until a predetermined decoding threshold is reached. By employing successive interference cancellation, AMA can achieve higher SE and accommodate more clients/users compared to conventional MA schemes. However, it also introduces additional complexity and requires accurate channel state information for efficient interference cancellation. AMA and SIC are active areas of research, and their performance depends on various factors such as the number of clients/users, channel conditions, interference level, and system parameters. Ongoing research aims to optimize the design and performance of AMA systems and explore their potential applications in future wireless communication networks.

Pattern division multiple access (PDMA) is a multiple access technique utilized in wireless communication networks. PDMA grants many clients/users to access the ditto frequency band concurrently by splitting the available bandwidth into a set of orthogonal patterns. These patterns are then assigned to different clients/users in such a way that they do not interfere with each other. In PDMA, each user is assigned a unique pattern from the pattern set, which is used to modulate the data transmitted by that user. The patterns are designed to be orthogonal, meaning that they have no correlation with each other, and therefore do not affect each other. This permits several clients/users to send and obtain data at the same time and on the same frequency band without causing any interference. PDMA has several advantages over other multiple access schemes, including high spectral efficiency, low interference, and scalability. However, it also has some limitations, such as high implementation complexity and high computational requirements. [11] describes a study that addressed the issues of poor detection performance and large complexity of traditional methods. Utilizing PDMA as an instance, [11] enhances the prevailing receiver detection algorithms such as MPA (message passing algorithm) and EPA (expectation propagation algorithm), MMSE-SIC (minimum mean square error-serial interference cancellation), and puts forward equivalent improved algorithms with specific formulas (i.e. turbo-like MPA (MPA-turbo)), enhanced-MMSE-SIC, and turbo-like EPA with hybrid parallel interference cancellation (EPA-Turbo-PIC) [12]. PDMA is still an emerging technology, and there is ongoing research to improve its performance and reduce its implementation complexity. It has the potential to show an important role in future wireless communication networks especially in scenarios where high spectral efficiency and low interference are critical, such as IoT, and 5G networks.

Advantages of Code Domain NOMA

Spectral efficiency: Code domain NOMA utilizes advanced coding techniques therefore this leads to higher SE compared to traditional OMA techniques.

Increased capacity: By enabling simultaneous transmission of many clients/users in the similar resource block, code domain NOMA increases the overall system capacity. This is particularly beneficial in consequences with many connected gadgets and limited frequency resources.

Fairness: Code domain NOMA can provide improved fairness among clients by allocating more power and resources to clients with poor channel conditions. This ensures that all clients/users have a reasonable QoS and reduces the occurrence of outage for clients with weak signal strength.

Low latency: The use of code domain NOMA can reduce latency in wireless communication systems. By allowing concurrent transmission and reception of multiple clients/users, the overall communication delay is decreased, enabling faster data exchange and response times.

Disadvantages of Code Domain NOMA

Complexity: Implementing code domain NOMA requires advanced signal processing techniques and complex receiver algorithms to separate and decode multiple user signals. This complexity adds to the computational and implementation overhead, requiring more powerful and sophisticated hardware.

Interference: Code domain NOMA relies on the use of interference cancellation techniques to separate user signals. However, in scenarios with strong interference, such as dense networks or congested environments, interference cancellation may become challenging, leading to performance degradation.

Channel conditions: Code domain NOMA performs better when clients/users experience significantly different channel conditions. If clients/users have similar channel characteristics, the benefits of NOMA may diminish, and traditional orthogonal techniques may be more suitable.

System design and optimization: Proper system design and optimization are crucial for realizing the benefits of code domain NOMA. This includes careful power allocation, user grouping, and resource allocation strategies. Designing an efficient NOMA system requires extensive modeling, optimization, and coordination among different network entities.

It's essential to notice that the advantages and disadvantages of code domain NOMA can vary depending on the specific deployment scenario and the system requirements.

6.4.5.2 Multiple Input Multiple Output Non-Orthogonal Multiple Access (MIMO-NOMA)

MIMO-NOMA is a WCT that uses many antennas on both the transmitter and receiver side, along with NOMA. This combination enhances the capacity and spectral efficiency of the systems by efficiently utilizing each antenna's signals. MIMO technology increases the wireless communication capacity and enhances its quality by using multiple antennas at both ends of transmission. MIMO technology can improve wireless signal quality and combat the effects of fading by utilizing multiple antennas to make use of spatial diversity in the channel. NOMA allows many clients/users to share the common subchannel by using power domain multiplexing, where each user is allocated a different power level, instead of a unique subchannel. MIMO NOMA combines these two technologies to provide even greater capacity and spectral efficiency. By using NOMA, many clients/users can give out common time and frequency resources, and by using MIMO, the system can exploit the spatial diversity of the wireless channel. The combination of these two technologies can result in a significant increase in system capacity, especially in high user density scenarios.

In downlink MIMO-NOMA systems, beamforming is employed to create B distinct beams in the spatial domain using M_{BS} BS antennas, as shown in Figure 6.7. Multiple clients/users' signals can be superimposed within a beam, leading to the thought of IBAM, which is akin to conventional NOMA employing SIC. The

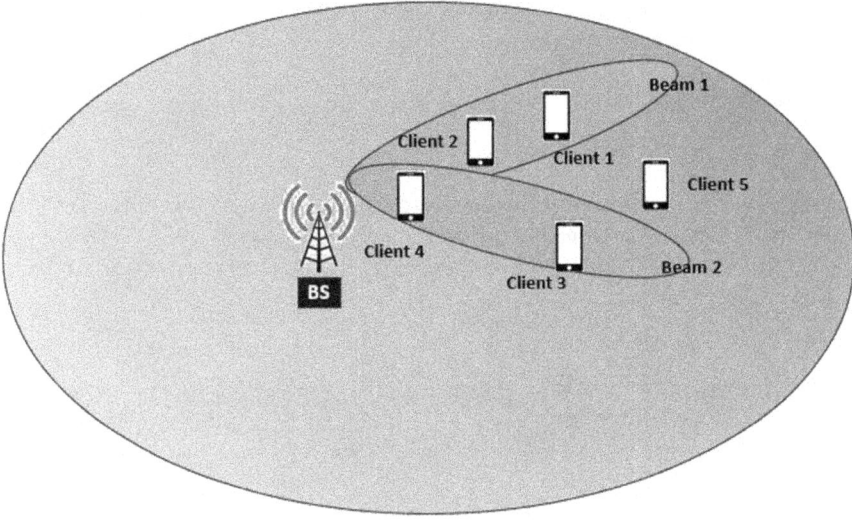

FIGURE 6.7   A scheme od downlink MIMO-NOMA.

transmitter beamforming vector for b^{th} ($1 \leq b \leq B$) is designated as m_b. Assume that k_b is the total number of clients/users in the bth beam, $x_{b,I}$ is the ith user's transmitted symbol in b^{th} beam, and $p_{b,i}$ is the associated power-scaling coefficient. Then, by adding up all the signals from the B distinct beams, the M_{BS}-dimensional vector of transmitted downlink signal at the BS can be written as [18]

$$x_0 = \sum_{b=1}^{B} m_b \sum_{i=1}^{k_b} \sqrt{P_{b,i}} x_{b,i}. \tag{6.17}$$

In the scenario that every user has Nr receiver antennas, the ith user's Nr-dimensional vector of received signal contained within the bth beam can be represented as

$$y_{b,i} = H_{b,i} x_0 + v_{b,i}, \tag{6.18}$$

where $v_{b,i}$ stands for the Gaussian noise plus the inter-cell interference and $H_{b,i}$ stands for the channel matrix of size $Nr \times M_{BS}$ between the ith user in the bth beam and the BS.

Inter-beam and intra-beam interference can be eliminated at the receiver by employing two different interference cancellation techniques. Spatial filtering, which is like the signal identification process employed in SDMA systems, can reduce inter-beam interference. Assuming that the ith user's spatial filtering vector in the bth beam is $f_{b,i}$, the signal $z_{b,i}$ after spatial filtering can be written as [18]

$$z_{b,i} = f_{b,i}^H y_{b,i}$$
$$= f_{b,i}^H m_b \sum_{j=1}^{k_b} \sqrt{P_{b,j}} x_{b,j} + f_{b,i}^H H_{b,i} \sum_{\substack{b'=1 \\ b' \neq b}}^{B} m_{b'} \sum_{j=1}^{k_b} \sqrt{P_{b',j}} x_{b',j} + f_{b,i}^H v_{b,i}. \tag{6.19}$$

By setting the total power of the ICI, receiver noise and inter-beam interference to unity, (6.19) can be written as

$$z_{b,i} = \sqrt{a_{b,i}} \sum_{j=1}^{k_b} \sqrt{P_{b,j}} x_{b,j} + q_{b,i}, \qquad (6.20)$$

where $q_{b,i}$ is the normalized term that represents the ICI, receiver noise and inter-beam interference added together, while $a_{b,i}$ is formulated as

$$a_{b,i} = \frac{\left| f_{b,i} H_{b,i} m_b \right|^2}{\left\{ \displaystyle\sum_{\substack{b'=1 \\ b' \neq b}}^{B} \sum_{j=1}^{k_b} P_{b',j} \left| f_{b,i} H_{b,i} m_{b'} \right|^2 + f_{b,i}^H E\left[v_{b,i} v_{b,i}^H \right] f_{b,i} \right\}}. \qquad (6.21)$$

The system model (6.20) closely mirrors the basic NOMA mixed with SIC outlined above. By implementing intra-beam SIC after spatial filtering, both inter-beam and inter-user interference caused by superposition coding within a single beam can be significantly reduced.

It is a natural outcome that when multiple users/clients give out a single beamforming vector, a greater number of users/clients can function at the same time. For more information, see Figure 6.8 [18], which shows that both the extended NOMA using MIMO and the basic NOMA mixed with SIC can reach a greater sum rate than OFDMA. Compared to the standard NOMA approach that uses only one antenna at the BS, using MIMO systems in NOMA can significantly increase the sum rate. Moreover, the number of reference signals required for this NOMA approach stays consistent with the number of antennas of transmitter, irrespective of non-orthogonal user signals present. This approach ensures that the number of downlink reference

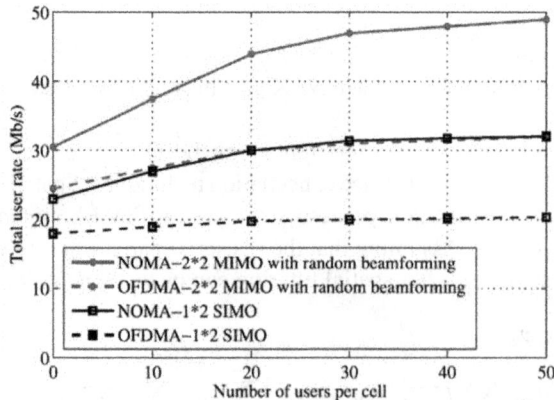

FIGURE 6.8 NOMA's opportunistic random beamforming downlink system performance.

signals stays within the limit of transmitter antennas when the number of clients/users exceeds it.

In the uplink systems, each user connects to K clients/users using a single transmitter antenna. The BS has M_{BS} antennas and can obtain signals that can be expressed as

$$y = \sum_{i=1}^{K} h_i \sqrt{P_i} x_i + V, \tag{6.22}$$

where h_i is the user i to BS MBS dimensional channel vector. Additionally, v stands for the combination of the inter-cell interference vector and the Gaussian noise, whereas x_i and p_i are the signal transmitted and the transmit power by the i-th user, respectively.

Overall, MIMO-NOMA is an emerging technology for coming wireless communication systems along with the advantage of increased spectral efficiency, high-user density, lower latency, but its deployment requires careful consideration of the trade-offs between complexity, interference management, power allocation, and coverage.

6.4.5.3   Cooperative NOMA

Cooperative NOMA is a communication technique where multiple clients/users work together to increase the overall system performance. In cooperative NOMA, clients/users share their resources, such as power and channel resources, to enhance the transmission of their data. In cooperative NOMA, clients/users are split into two groups: the clients/users that act as relays (or helpers), and the clients/users that act as the primary transmitters. The relays assist the primary transmitters by amplifying and forwarding their signals to the intended receivers, using the NOMA transmission scheme. Like the standard NOMA, cooperative NOMA also detects the multi-user signal using a SIC receiver.

To improve the dependability of reception for clients/users who experience poor channel conditions, we can use other clients/users who have better channels as relays. Cooperative transmission differs from NOMA with SIC in its ability to send signals from clients/users with strong channel conditions to those with weaker ones using short-range communication methods like Bluetooth and ultra-wide band (UWB) techniques. Figure 6.9 depicts a lossless cooperative NOMA downlink system model supporting L clients/users and relaying on a single BS according to their channel conditions [13]. The following two phases are essential to cooperative NOMA [15].

Direct transmission is represented by the first phase, often known as the broadcast phase. The BS distributes downlink or downstream messages to all L clients/users during this phase using the basic NOMA SIC principle, where the L clients/users' superimposed information abides by the overall power constraint. The SIC process is put into action on the user side, which results in clients/users with weaker network conditions being able to receive signals intended for those with stronger ones [16].

During the cooperative transmission second phase, participating clients/users utilize short-range communication channels like Bluetooth or UWB to share their

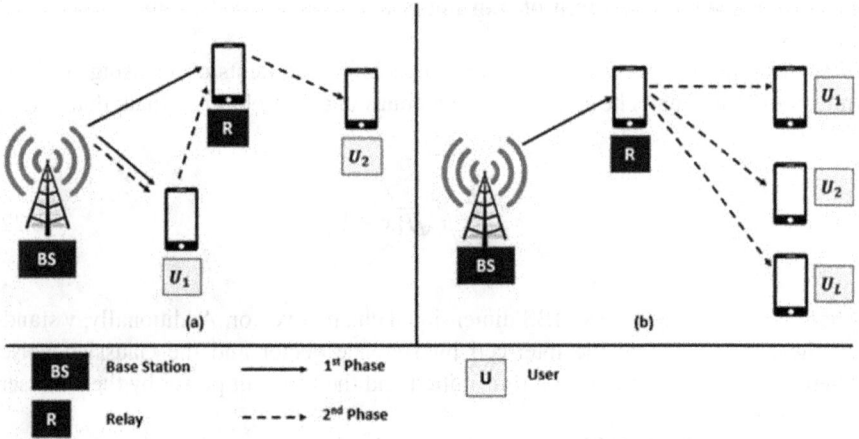

FIGURE 6.9 Cooperative NOMA downlink system model. (a) Coordination of transmission via direct and relay. (b) A cooperative plan with no underlying connection.

signals. This phase has (L-1) time slots, with the Lth user transmitting a superposition of (L-1) signals intended for other clients/users during the first slot. Next, the SIC procedure is initiated another time at these (L-1) users/clients and the $(L-1)^{th}$ user/ clients gets access to its own data at a higher SNR using maximum ratio combining (MRC), which combines all signals received across both phases.

Cooperative NOMA can improve the system performance by reducing interference and improving the coverage area. By using cooperative NOMA, the signals transmitted by primary clients/users can be boosted by the relays, which can increase the SNR at the receivers. This, in turn, can reduce the probability of error and increase the coverage area of the system. Furthermore, cooperative NOMA can also enhance the system capacity by utilizing the multiuser diversity. In cooperative NOMA, clients/users with stronger signals can act as intermediaries to assist clients/users with weaker signals, thus increasing the overall system throughput. The SVM (support vector machine), one of the commonly used supervised learning algorithms, has demonstrated optimistic results in relay selection. Recently, a relay selection is treated as a support vector machine (SVM) multi-class classification issue in work [14] proposed by the author. The primary idea behind the SVM classification algorithm is to train the classifiers using many system realizations to optimize their parameters. The main benefit of this approach is that the training phase may be completed offline because optimizing the SVM classifier parameters needs a lot of computing power. The trained SVM multi-class classifiers can then be employed immediately during system operation with no further training needed. Recently, an analysis was conducted on the power distribution and decoding order at the BS of cooperative NOMA (C-NOMA)-based two-user uplink (UL) cellular networks. This analysis was published in another work [16]. The least user attainable rate is the goal of the joint optimization problem that the authors formulate. As an alternative approach, authors develop simplified and highly effective analytical formulas to determine power allocation coefficients, even when dealing with non-convex problems.

Improved spectral efficiency, enhanced energy efficiency, better coverage, and increased fairness are some advantages of cooperative NOMA and increased complexity, high latency, interference, and security concerns are some disadvantages of cooperative NOMA.

6.4.5.4 Bit Division Multiplexing (BDM)

BDM's foundation is hierarchical modulation [17], but it distributes resources at the bit level instead of the symbol level. Even though multiuser signals may share a similar constellation, indicating overlay in the symbol domain, BDM's resource distribution is technically orthogonal in the bit domain.

6.4.5.5 Multi-user Shared Access (MUSA)

Another code-domain multiplexing-based NOMA scheme that can be thought of as an enhanced CDMA-style scheme is MUSA.

Symbols transmitted by a user in the MUSA system shown in Figure 6.10 are multiplied together in the uplink stage. Different spreading sequences can be assigned to each symbol from the same user, resulting in interference averaging advantages. The symbols are then spread over the same time-frequency resources like OFDM subcarriers and sent out as a whole. Without sacrificing generality, we assume that there are K clients/users and N subcarriers, and that each user always sends a single symbol. MUSA can also support rank-deficient conditions (i.e., $K > N$),

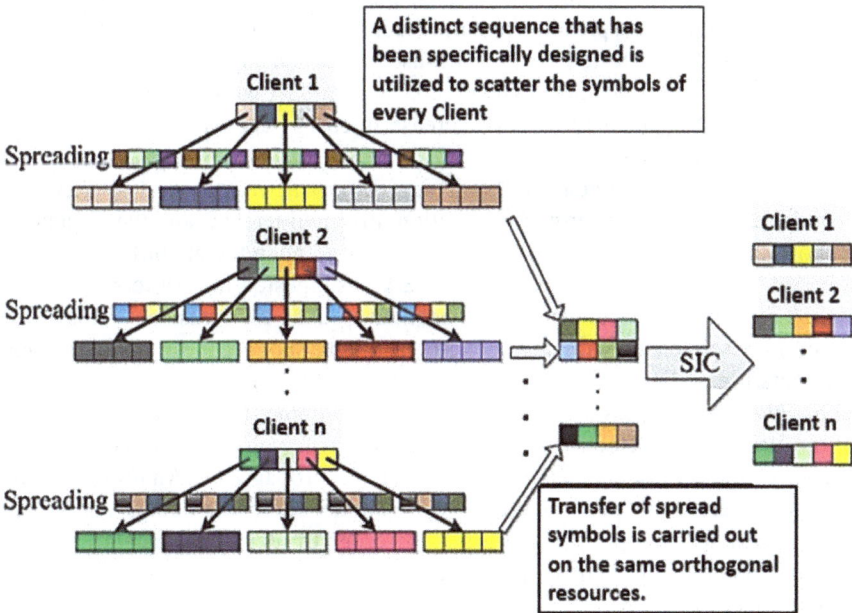

FIGURE 6.10 MUSA system for uplink.

which will cause user interference. Data from the several clients/users are separated at the receiver using SIC and linear processing, depending on the channel circumstances of each user.

The clients/users of the MUSA system are divided into G groups for the downlink. The various user symbols in each group are then superimposed after being weighted by various power-scaling coefficients. The symbols from G groups can be spread out by using orthogonal sequences of length G. This spreading technique allows the superimposed symbols to be propagated effectively. Clients/users belonging to the same group use identical spreading sequences, which are orthogonal to those of other groups. This technique eliminates inter-group interferences at the receiver. After this step, intra-group interference cancellation can be achieved with SIC by utilizing the resulting power differential.

Low cross-correlation spreading sequences in MUSA are required for nearly complete interference cancellation at the receiver. The accompanying SINR difference and SIC of the MUSA approach have the express benefit of increasing downlink capacity. Another strong advantage of MUSA is its ability to ensure fairness among the multiplexed clients/users while maintaining capacity. Proximal device discovery (DD) is important in wireless self-systematizing and network-aided D2D networks. It serves as the basis for self-governance, the expansion of the service area, and performance enhancements including multi-hop and cooperative transmission. For proximal DD, energy efficiency and discovery precision are crucial performance indicators.

For network-assisted D2D communication, a unique DD process based on NOMA and MUSA is proposed in [19]. To support dense device detection and improve detection distance in the mobile systems, the suggested shared access approach employs a complicated spreading pattern. Additionally, a shared access approach for discovery signals with additional multi-user detection algorithms is created. In comparison to the conventional DD approaches, MUSA with its spreading pattern demodulation produce a 12 percent increase in energy economy and a 98 percent improvement in total accuracy.

IoT is widely anticipated to play a significant role in 5G wireless networks. Low device costs, less energy consumption, minimal latency, and the capacity to support numerous simultaneous connections are some of the major IoT problems. A brand-new NOMA scheme called MUSA is suggested in [20] to support IoT. MUSA utilizes an advanced technique that spreads transmission across a set of complex sequences with short length, allowing support for a greater number of clients/users/clients who access the common radio resources. In combination with a grant-free access approach, this enables reliable interference cancellation on the BS side and effective handling of high traffic volume. A brand-new class of codes called 5-ary codes is utilized in [21] for propagation. An uplink MUSA system's bit error rate performance study is looked at for 5-ary codes and several short-length spreading sequence types. In order to identify clients/users using the SIC method, the analysis is also carried out when the ordering method is changed.

There are several advantages and disadvantages to MUSA.

Advantages

1. **Collaboration:** MUSA enables clients/users to work together on a project or share resources, fostering collaboration and teamwork.
2. **Efficiency:** With MUSA, clients/users can access shared resources or systems without the need to duplicate data or information, leading to increased efficiency and reduced redundancy.
3. **Cost savings:** MUSA can often be more cost-effective than providing each user with their own individual resources or systems.
4. **Resource optimization:** MUSA can help to optimize the use of resources, such as computer networks or machinery, by ensuring that they are being utilized to their maximum capacity.
5. **Accessibility:** MUSA can make resources more accessible to a larger number of clients/users, regardless of their location.

Disadvantages

1. **Security risks:** MUSA can increase the risk of security breaches, as multiple clients/users may have access to sensitive information or systems.
2. **Limited capacity:** In some cases, MUSA may be limited by the capacity or capabilities of the shared resource or system.
3. **Conflicts:** MUSA can sometimes lead to conflicts or collisions between clients/users, particularly if they have different priorities or needs.
4. **Complexity:** Implementing and managing MUSA can be complex, requiring careful planning and coordination.
5. **Maintenance:** MUSA requires ongoing maintenance and management to ensure that it remains functional and effective.

Overall, MUSA can be a powerful tool for promoting collaboration, efficiency, and resource optimization, but it also requires careful consideration of the potential risks and challenges involved.

6.5   RATE SPLITTING MULTIPLE ACCESS (RSMA)

The latest advancements in mobile technology, including "mTC (massive machine type communications)," URLLC (ultra reliable low latency communications)," and "eMBB (enhanced mobile broadband) are posing unique challenges to the development of 6G technology [22].

The difficulty of creating complex MA schemes is one of these challenges. MA refers to a broad collection of approaches that permit multiple clients/users. Cellular radio networks have evolved over time. In earlier versions, systems like orthogonal MA (OMA) were used to avoid interference by distributing unique radio resources to each user in the domains frequency, time, code and OFDMA in 1G through 5G.

The goal of minimizing inter-user interference and excessive transceiver complexity inspired the adoption of orthogonal radio resource allocation. An ineffective use of radio resources results from such a strategy, nevertheless. In addition to

OFDMA, 4G and 5G have taken advantage of numerous antennas such as in MIMO and massive MIMO (MaMIMO), to serve many users/clients in a non-orthogonal manner. Current MA schemes make use of various approaches for managing interference, including several methods that have been adopted previously such as

Interference considered as noise (like in SDMA)
Decoding interference (like in NOMA)

To achieve a more reliable 6G wireless connection with improved performance, innovative techniques must move away from conventional extreme interference management strategies such as SDMA (interference considered as noise) and NOMA (fully interference decoding). This shift can help to ensure reduced latency and higher data rates [23]. More general and powerful techniques which consider NOMA and SDMA as special cases have been developed recently that rely on linearly precoded rate splitting, termed as RSMA.

RSMA splits the message into separate components in both power and spatial domains, transmitting them simultaneously within the same time-frequency block, employing techniques from both PD-NOMA and SDMA. RSMA controls interference by altering the message split and power distribution. RSMA can maximize available resources, including frequency, time, spatial domains, and power to provide excellent degree of freedom (DoF) in the downlink. This feature works consistently well with both imperfect and perfect channel state information at the transmitter (CSIT).

RSMA comes up with adaptation between SDMA and NOMA, which leads to rate enhancement with lower computational capacity. Modifying message split and altering power allocation to private and common messages has capability to transit between the two schemes NOMA and SDMA [24] as shown in Figure 6.11.

RSMA has been reviewed in various ways depending on how clients/users' messages are divided and blended into common messages, as classified in Figure 6.12.

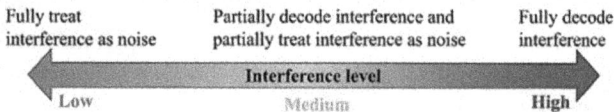

Fully treat interference as noise	Partially decode interference and partially treat interference as noise	Fully decode interference
	Interference level	
Low	Medium	High

TABLE
COMPARISON OF PREFERABLE INTERFERENCE LEVELS
OF DIFFERENT MA SCHEMES

Interference levels	Low	Medium	High
SDMA	√	×	×
NOMA	×	×	√
RSMA	√	√	√

Notations: √: Suited. ×: Not well suited.

FIGURE 6.11   A contrast of desirable interference levels between medium access schemes.

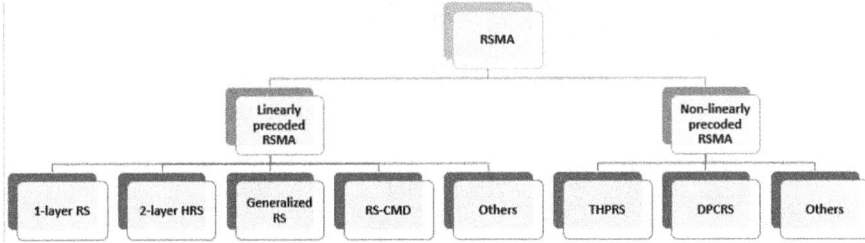

FIGURE 6.12   RSMA classification.

For understanding purposes, let us assume there are k single antenna clients/ users that are served by a base station having N transmitting antennas, then we would split the message of each client/user in two parts: one is private and another one is a common part. The common parts of all clients/users are combined to create a single shared message, which is then encoded into a unified stream as illustrated in Figure 6.13 [25].

Private information of clients/users is protected by encoding them into separate private streams, distinct from the rest. A total of k+1 streams are transmitted in 1-layer RSMA. Linear precoders are used to precode these streams. The transmit signal x is represented as

$$x = p_c \, s_c + \sum p_p \, s_p.$$ (6.23)

h_k is channel between user k and BS, z_k is AWGN at user k. The received signal at user k is given by

$$y_k = h_k^H x + z_k.$$ (6.24)

The common stream is first decoded by all clients/users at the receiver side while they consider all private streams as noise. Clients/users who correspond with each other can decode their own private stream by treating others' streams as background noise [25].

For simplicity, let us assume that each user has just 6 bits of data to send. Private and common stream-based transmission rates for both clients/users can be provided by 1bps/Hz and 0.5bps/Hz, respectively. These rates are calculated from the CSIT available at transmitter. Amount of message bits to be transferred is quadrupled in terms of the transmission rates. For example, the common stream message bits will be 0.5*4=2 bits and for private stream the number of message bits are 4 as given in Figure 6.14.

Figure 6.14 describes how the data bits are being divided before transmission. This assumption is far from reality, but is sufficient to understand the basic split of RSMA. The authors of [26] studied how effective RSMA is for an uplink/upstream communication system with an FBL. Their research focused on the impact of varying blocklength and target rate on the error probability performance and throughput of RSMA in a Single-Input Single—Output (SISO) Multiple Access Channel (MAC) with two user messages. The results demonstrate that, unlike NOMA, RSMA can enhance both throughput and error probability performance. It is a difficult but

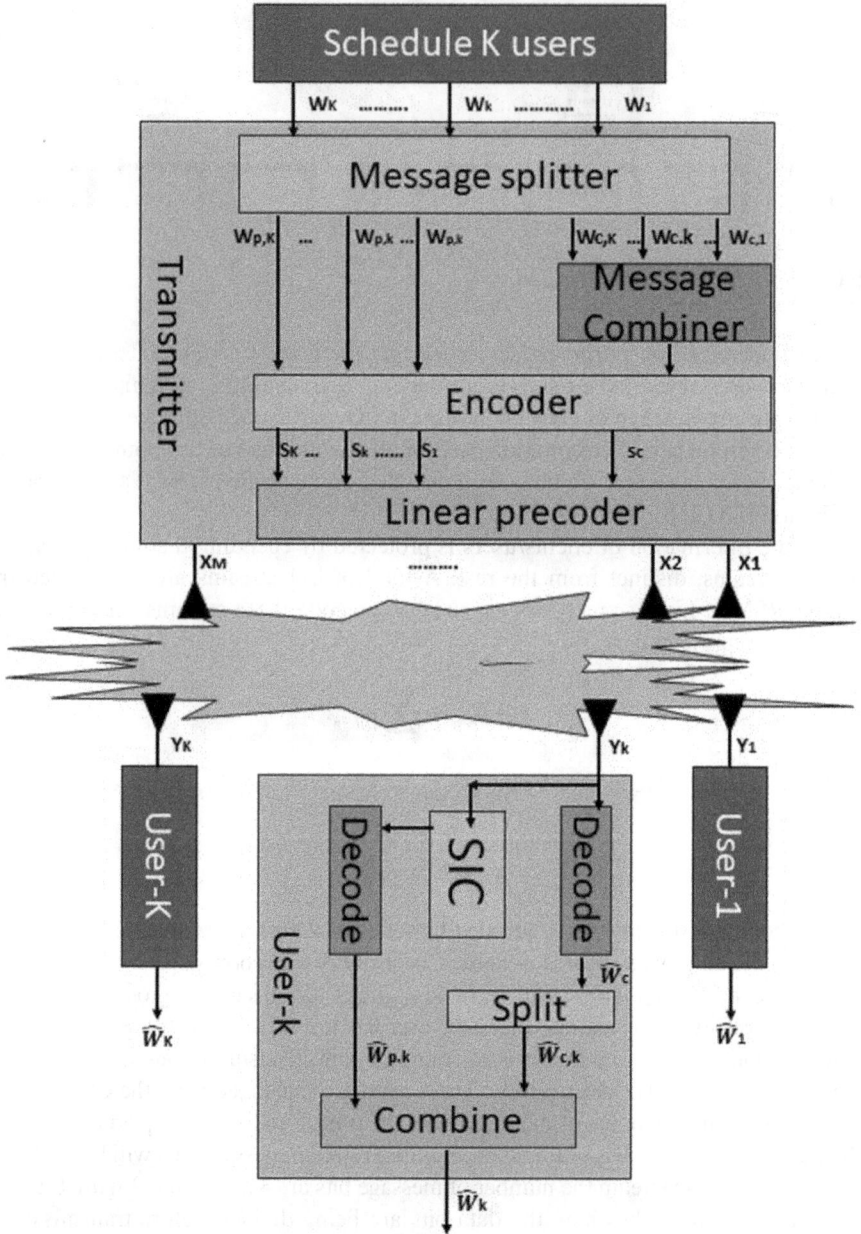

FIGURE 6.13 RSMA transceiver architecture.

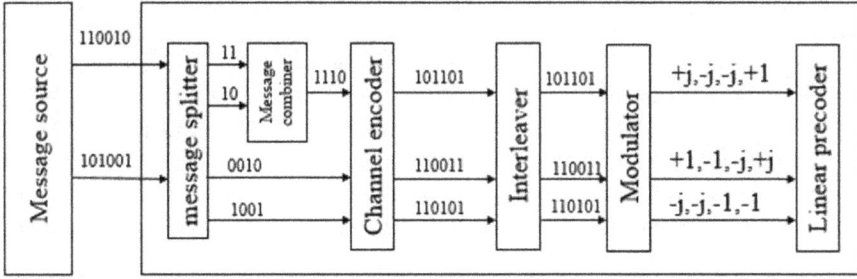

FIGURE 6.14   A 6-bit data transmission in RSMA.

important problem to find a linear precoding scheme that maximizes the "sum spectral efficiency" of RSMA. In their paper [27], the authors introduce a unique method for designing precoders that can simultaneously handle both shared and private messages using RSMA. First, they use a LogSumExp technique to roughly represent the non-smooth minimum function portion of the RSMA's sum spectral efficiency. Then, to make the sum spectral efficiency maximization issue tractable, writers reformulate it as a version of the log-sum of Rayleigh quotients. The research explains that breaking down the first-order optimality condition of the reformulated problem into an eigenvector-dependent nonlinear eigenvalue problem helps depict how the leading eigenvector of the optimality condition shows a local optimal solution [27]. Authors suggest an approach for locating the leading eigenvector that is modelled after a power iteration. The authors of [28] take into account both the imperfect "channel state information (CSI)" at the "transmitter (CSIT)" and the imperfect CSI at the receiver (CSIR) and present a max-min fairness optimization framework depending on RSMA to ensure each user's minimum level of QoS.

Most of the published research on RSMA make the assumptions that the blocklength is limitless and that all clients/users employ SIC to decode the shared stream. The user with the weakest channel quality is the one who limits the data rate of the shared stream according to the first supposition. In real-world systems with finite blocklengths, the second assumption could result in less-than-ideal performance. The authors of [29] suggest a versatile RSMA approach that enables the system to choose whether or not a user should employ SIC to decode the shared stream. For beamforming design, writers create a low-complexity method as well as an ideal algorithm. They come up with a useful approach for the selection of RS-clients/users and construct a semi-closed-form solution for the optimal data rates [29].

Tera-hertz massive MIMO communication systems can have their service coverage greatly improved by reconfigurable intelligent surfaces (RIS). The performance of traditional "spatial division multiple access (SDMA)" is significantly diminished by the difficulty in collecting correct high-dimensional CSI with little pilot and feedback signaling overhead. For "RIS-aided tera-hertz multi-user MIMO" systems, [30] presents a deep learning (DL)-RSMA technique to increase robustness against CSI imperfection. The authors proposed "a new hybrid data-model driven DL-based

RSMA precoding scheme for improving wireless communication." It involves passive precoding at the RIS, analogue active precoding, and RSMA digital active precoding performed at the BS. The authors suggest a "transformer-based data-driven RIS reflecting network (RRN)" to provide passive precoding at the RIS. The "LoS-MIMO antenna array architecture" is used by both BS and RIS, which recommends an analogue precoding technique based on a match-filter for the analogue active precoding at the BS.

The authors propose a simpler scheme called approximate weighted minimum mean square error (AWMMSE) for digital precoding at the BS. They use this newly proposed scheme to create a deeper and more efficient network called deep unfolding active precoding network (DFAPN) by incorporating DL techniques. Then, we provide a "CSI acquisition network (CAN)" with minimal pilot and feedback signaling overhead to acquire precise CSI at the BS for the examined RSMA precoding technique to obtain improved spectrum efficiency. The proposed RSMA scheme that utilizes DL shows promise in enhancing the robustness against CSI imperfections, leading to improved SE and less signaling overhead [30].

There are several advantages and disadvantages of RSMA.

Advantages
1. **Increased spectral efficiency:** RSMA makes it possible for more data to be transmitted over the same bandwidth, which is particularly useful in wireless networks where bandwidth is often limited.
2. **Low latency:** RSMA can achieve lower latency compared to other MA techniques, such as TDMA and FDMA, because each user can transmit their data simultaneously.
3. **Flexible resource allocation:** RSMA allows for flexible resource allocation, as each user can allocate its resources based on its own needs. This means that the system can adapt to changing traffic demands and provide efficient resource utilization.

Disadvantages
1. **Complexity:** RSMA requires complex encoding and decoding algorithms to ensure that each user's data can be transmitted simultaneously. This increases the complexity and may require more processing power.
2. **Interference:** RSMA can suffer from interference between clients/users, which can lead to decreased performance. This is particularly true when the number of clients/users sharing the same frequency band increases.
3. **Limited scalability:** As more clients/users join the system, the complexity of RSMA tends to increase, which may make it difficult for the system to be easily scalable for a larger number of clients/users.

6.6 RATELESS MULTIPLE ACCESS (RMA)

RMA is a communication protocol that enables multiple clients/users to access a shared communication channel in a decentralized manner without requiring explicit coordination or scheduling. In RMA, clients/users transmit packets randomly and independently of each other, and the receiver uses a decoding algorithm to recover the transmitted data. RMA has a unique characteristic that allows for data recovery by the receiver, even in cases where the number of clients/users is variable and unknown over time.

RMA achieves its resilience to changes in the number of clients/users by using a rateless code. A rateless code is a code that can generate an infinite number of encoded packets from a finite set of source packets. RMA encodes the data packets using a rateless code and transmits the encoded packets over the communication channel. The recipient accumulates a group of encoded data packages and applies a decoding algorithm to regain the initial data packets. By utilizing a rateless code, the recipient can recover the initial data packets even if they receive fewer encoded packets than the entire number of encoded packets transmitted.

RMA has several advantages over other MA schemes, such as achieving concurrent high reliability, random access, less latency, scheduled access and enormous connectivity. RMA can handle a dynamic number of clients/users without requiring explicit coordination or scheduling, making it suitable for ad hoc networks and IoT devices. RMA also has a low overhead, as clients/users can transmit packets randomly and independently, without requiring any coordination. Additionally, RMA is resilient to packet loss and interference, as the receiver can recover the data packets even if some of the encoded packets are lost.

In a recent study [31], researchers examined the effectiveness of the RMA scheme's maximum likelihood (ML) decoding using BPSK" modulation in an "AWGN channel. The RMA system's ensemble weight distribution was initially established by [31]. When subjected to BPSK modulation in an AWGN channel, the authors determined the highest possible limit on the RMA system's decoding error performance.

Here are some advantages and disadvantages of using RMA.

Advantages

1. **Improved efficiency:** RMA can significantly increase the efficiency of WCN by allowing many clients/users to share the same channel without collisions. This is because RMA is designed to allocate the channel resources dynamically, based on the user's channel conditions and transmission rate.
2. **Increased reliability:** RMA can improve the reliability of wireless communication by providing a mechanism for retransmission of lost packets. RMA uses a rateless coding scheme to encode the data, which means that the receiver can decode the data even if some packets are lost during transmission.
3. **Flexibility:** RMA is a flexible technique that can be used in a wide range of wireless communication scenarios, including point-to-point, multi-hop, and broadcast communication.

Disadvantages

1. **Increased complexity:** RMA requires more complex encoding and decoding algorithms than traditional wireless communication techniques, which can increase the processing power and memory requirements of the communication devices.

2. **Limited scalability:** RMA may not be scalable to large-scale wireless networks. This means that the people who can use a particular channel at the same time is limited.
3. **Increased overhead:** RMA may introduce additional overhead in the wireless communication system, as it requires the transmission of additional coding symbols for error correction and retransmission of lost packets. This can reduce the overall throughput of the system.

Overall, RMA is a promising MA scheme that offers several advantages over other schemes. It is particularly suitable for decentralized networks and IoT devices that require low overhead and resilience to changes in the number of clients/users.

6.7　COORDINATED MULTI-POINT-NOMA (CoMP-NOMA)

CoMP-NOMA is a communication technique that combines two key concepts: coordinated multi-point (CoMP) and NOMA. CoMP involves coordinating the transmission and reception of data across multiple base stations (BSs) in a cellular network. This coordination allows for improved signal quality, increased capacity, and enhanced coverage by mitigating interference and optimizing resource allocation. By using NOMA, multiple people can be served simultaneously on the same time-frequency resources. This is a way to increase spectral efficiency, which means that more information can be sent in less time. CoMP techniques are employed to coordinate the transmissions between BSs and clients/users, reducing interference and improving system-wide performance.

Figure 6.15 shows a NOMA model for downlink. Here the signals broadcast from BS 2 may cause user 1 to experience major interference, and user 3 may experience severe interference brought on by BS1 transmitted signal. As a result, the NOMA in CoMP scheme may perform noticeably worse.

The signals of all NOMA clients/users can be used in cooperative transmit-precoding to reduce inter-cell interference in the downlink. The process of finding the best transmit precoder isn't easy. To achieve this, all user signals and conditions of channels must be accessible to the concerned BSs. However, in situations where geographically distant BS antennas produce a beam that fails to cover multiple angularly separated clients/users for intra-beam NOMA in CoMP situations may be ineffective.

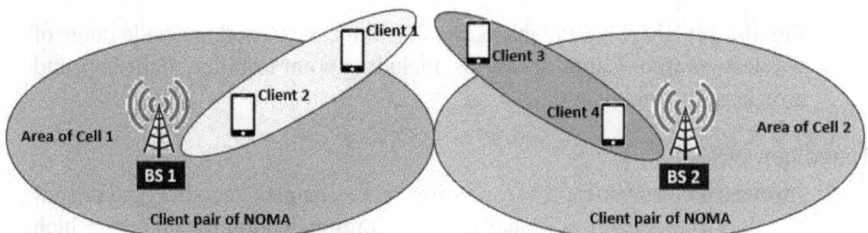

FIGURE 6.15　CoMP using NOMA.

A new precoding scheme was suggested to simplify transmission in a NOMA-CoMP setting [32]. The scheme uses the varying channel impulse responses (CIRs) of multiple clients/users to target signals seemingly found only at cell edges. Specifically, the scheme applies only to signals from edge clients/users, like those identified as user 1 and 3 in Figure 6.15.

In [33], researchers considered a multi-tier uplink NOMA system, analyzing the coverage probability of clients/users based on their rank with respect to distance from the serving stations. They examined the average rate coverage probability and the rate coverage probability for a user.

It has been demonstrated that at larger user densities per cell and higher target rate needs, a NOMA cluster's average rate coverage is superior than that of its equivalent OMA cluster. Additionally, [34] provides a current literature analysis of interference management approaches that use NOMA in multi-cell networks, covering both NOMA that relies on coordinated scheduling/beamforming and NOMA that uses joint processing. Aside from SIC implementation concerns, flawed CSI, multi-user power allocation, and clustering, [34] also highlights the key practical obstacles and difficulties that emerge in the implementation of multi-cell NOMA.

For a network with virtualized various infrastructures, writers in [35] develop a generalized joint transmission coordinated multi-point (JT-CoMP)-non-orthogonal multiple access (NOMA) model. Wireless network virtualization (WNV) allows for multiple joint broadcasts of CoMP in multi-infrastructure networks, providing benefits to all clients/users within this architecture. The CoMP-NOMA protocol generates an unlimited NOMA clustering (UNC) scheme that exhibits the highest attainable level for each NOMA cluster. The authors demonstrate that UNC causes consumers to have the highest level of effective interference cancellation (SIC) complexity. In this context, they suggest a limited NOMA clustering (LNC) strategy, in which only a portion of clients/users are subjected to SIC. According to numerical analyses, WNV and LNC lower user SIC complexity by up to 46 percent and 35 percent, respectively, and enhance user sum rate compared to the non-virtualized CoMP-NOMA system and UNC model. Since open and intelligent radio access networks are a good fit for their implementation, the suggested algorithms [35] are a good choice.

Advantages of CoMP-NOMA

1. **Increased SE:** CoMP-NOMA allows for many clients/users to share the same time-frequency resources, resulting in higher SE. This leads to increased capacity and improved utilization of network resources.
2. **Enhanced coverage and QoS:** CoMP-NOMA leverages coordinated transmission between BSs to mitigate interference and optimize resource allocation. This coordination improves coverage and QoS by reducing signal degradation and ensuring reliable communication in areas with weak signal strength.
3. **Improved system performance:** The combination of CoMP and NOMA techniques in CoMP-NOMA improves overall system performance. It enables better load balancing, reduces congestion, and enhances network capacity, resulting in a more efficient and robust cellular network.

Disadvantages of CoMP-NOMA

1. **Complexity and overhead:** Implementing CoMP-NOMA requires additional coordination and signaling overhead between multiple BSs. This can increase the complexity of network management and introduce additional latency in the system.
2. **Interference management challenges:** CoMP-NOMA relies on effective interference management techniques to mitigate co-channel interference. Achieving optimal interference coordination and resource allocation across multiple BSs and clients/users can be challenging and may require sophisticated algorithms and high computational resources.
3. **Compatibility and interoperability:** CoMP-NOMA may require modifications to existing network infrastructure and user devices to support the required coordination and NOMA. Ensuring compatibility and interoperability between different generations of network equipment and user devices can be a challenge during the transition to CoMP-NOMA.

It is important to mention that the advantages and disadvantages of CoMP-NOMA can differ according to specific factors like implementation, network environment, and deployment scenarios. The future scope of CoMP-NOMA holds several potential advancements and applications in the field of wireless communication such as 5G and beyond, massive IoT, ultra-dense networks (UDN), cooperative edge computing, millimeter-wave (mm-wave) communications, cooperative and cognitive networks, AI, ML.

6.8 NOMA-ORTHOGONAL TIME FREQUENCY SPACE (NOMA-OTFS)

Researchers across the world are eagerly exploring OTFS [36]. OTFS has demonstrated outstanding performance improvements in high mobility scenarios [37–40]. According to [41] and [40], OTFS and OMA-OTFS have been mentioned, whereas [42] reports that PD-NOMA techniques show better sum spectral efficiency (SE) performance compared to OMA and code-division NOMA schemes. This improvement in performance has been noted in [43] and [44]. The authors of [45] proposed NOMA with OTFS. The article [44] delves into beamforming in OTFS assisted NOMA networks with multiantenna BSs. The aim is to optimize data rate for low-mobility NOMA clients/users and high mobility OTFS clients/ users [46].

The authors of [47] have introduced a novel NOMA-OTFS system that showcases promising results both at the link-level and system-level. The challenge with combining OTFS transmission technology and NOMA lies in their different strengths. While OTFS excels in high-mobility scenarios, NOMA delivers better spectral efficiency in low-mobility situations compared to conventional OMA.

The researchers developed an minimum mean square error (MMSE)-SIC-based receiver for NOMA-OTFS for maximizing the SE and mobility. They have improved SINR for enhancing performance of the network.

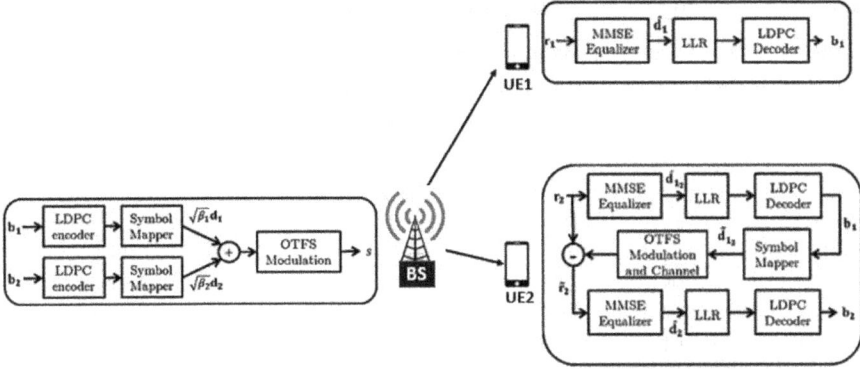

FIGURE 6.16 Two-user NOMA-OTFS system block diagram in downlink.

The SIC OTFS-NOMA receiver for downlink transmission, as illustrated by Figure 6.16 [47] is implemented practically for two clients/users. The BS creates a data stream for both the clients/users (represented by b1 and b2, respectively), encodes it using an LDPC encoder, and subsequently modulates massage. Figure 6.16 denotes the received signals for clients/users as d1 and d2, respectively. The modulated symbols are then superimposed with assigned power (βi). The time domain signal, overlaid with extra information, gets a boost through modulation for OTFS using the SFFT and Heisenberg transformation methods. These resulting signals are then transmitted through the delay-Doppler channel to multiple clients/users.

Figure 6.17 [43] highlights the effectiveness of using an OTFS-NOMA together using LDPC. The results shown in [47] outperform traditional OMA-OTFS. The future scope of OTFS-NOMA holds several potential advancements and applications in the field of wireless communication such as 5G and beyond, massive IoT, UDN, edge computing, mm-wave communications, cognitive networks, and AI.

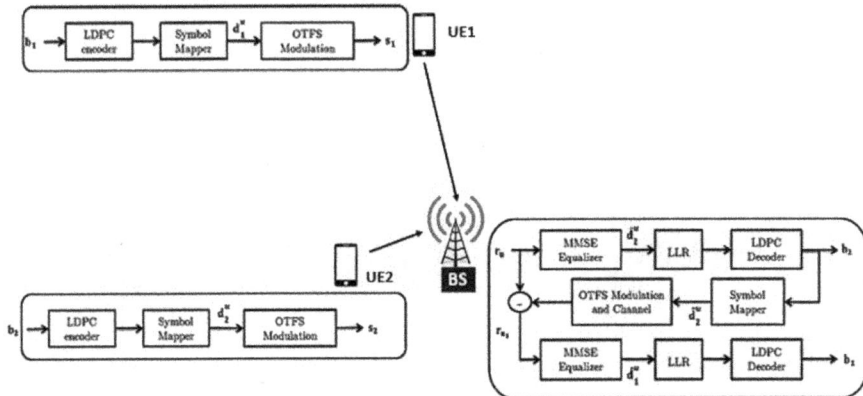

FIGURE 6.17 Two-subscriber NOMA-OTFS for uplink.

6.9 DELTA-ORTHOGONAL MULTIPLE ACCESS (D-OMA)

D-OMA is a new scheme enabling multiple devices to share the same channel without interfering with each other. D-OMA was proposed as an alternative to the traditional OFDMA technology, which is widely used in 4G and 5G wireless networks.

The challenge of spectrum scarcity can be addressed by adopting higher frequency bands such as mm-wave and beyond. However, these bands may not always be the best solution, particularly for medium to large communication ranges since they require considerable beam-steering skills due to extremely high attenuation.

Fully adopting OMA schemes within the available spectrum won't suffice for MA, whereas relying solely on NOMA methods lacks the flexibility to support wireless connectivity for devices with differing service requirements [48]. New techniques must be developed to manage interference and allocate resources for the cell-less networks due to their limited spectrum resources.

Deploying in-band NOMA clusters on a large scale has been impeded due to two major obstacles: the high complexity and increased power consumption at terminal devices. However, a new massive multiple-access scheme called D-OMA, proposed by the authors of [49], offers an effective solution to these problems. As such, it is considered suitable for 6G network architectures and requirements (Figure 6.18 (c)) [49].

In D-OMA, we allow for overlapping of different NOMA clusters with adjacent frequency bands by δ percnet of their maximum allocated subband. A δ value of 0 represents the conventional power-domain NOMA approach. To address overlap between clusters in NOMA, the size of each cluster may be reduced. This maintains INI and ICI counts similar to those found in non-overlapping massive in-band NOMA clustering.

By utilizing additional clusters within the same bandwidth allocation, it is possible to maintain the spectral efficiency achieved by a massive in-band NOMA cluster. This holds true for B Hz in Figure 6.18 (a and c). Reducing the size of different NOMA clusters can significantly decrease the complexity requirements and

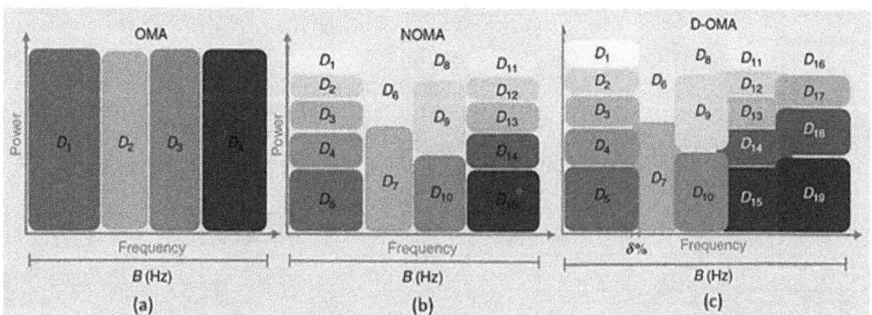

FIGURE 6.18   An example of the (a) OMA, (b) NOMA, and (c) D-OMA schemes.

power consumption for various devices without sacrificing performance standards. Optimizing NOMA clusters and power allocation involves designing the percentage of overlap among the clusters (δ). This is a crucial factor in achieving optimal performance.

The authors of [50] examined how well D-OMA performs in the uplink and created a multi-objective optimization plan to increase uplink EE in a multi-access point (AP) network empowered by D-OMA. They focused on enhancing the sub-channel and transmiting power allocations as well as the overlapping spectrum percentage between adjacent subchannels.

The study involving [50] participants concluded that D-OMA outperforms conventional NOMA and OFDMA when joint optimization of adjacent sub-channel overlap and scheduling is implemented.

D-OMA has several advantages over OFDMA and NOMA, including lower latency, higher throughput, and better scalability. However, it also has some disadvantages, such as higher complexity and the need for precise synchronization between the transmitter and receiver. Here are some advantages and disadvantages of using D-OMA given below.

Advantages of D-OMA

1. **High SE:** D-OMA can achieve higher SE compared to traditional MA schemes. This is because it allows multiple clients/users to share the same frequency band without causing significant interference.
2. **Low latency:** D-OMA is designed to give low latency, which is critical for real-time applications such as video conferencing, gaming, and virtual reality.
3. **Robustness:** D-OMA is more robust to interference and noise compared to traditional MA schemes. This is because it uses advanced signal processing techniques to separate different clients/users' signals.

Disadvantages of D-OMA

1. **Complexity:** D-OMA is more complex compared to traditional MA schemes. This is because it requires advanced signal processing algorithms to separate different clients/users' signals.
2. **Limited deployment:** D-OMA is a relatively new technology, and it is not widely deployed yet. This means that it may not be compatible with existing wireless communication systems.
3. **Compatibility issues:** D-OMA may not be compatible with some existing wireless communication systems, which could lead to interoperability issues.

D-OMA is still an emerging technology, and further research is needed to fully understand its capabilities and limitations. D-OMA gives better performance than OFDMA and NOMA but there are still some implementation issues and open research challenges such as method of power allocation, clustering of devices/users, requirement of an efficient SIC method, and spectrum variations.

6.10   VISIBLE LIGHT COMMUNICATION-NOMA (VLC-NOMA)

VLC was recently proposed as a reliable replacement for indoor wireless communication because of increased bandwidth demands, its energy efficiency, and its ability to deliver ubiquitous connection [51] that do not meet by the RF spectrum.

Scholarly articles have reported on the utilization of NOMA techniques in VLC, with numerous reasons making it an optimal choice for downlink systems in this context, as outlined in [52–53].

NOMA technology can function properly with just a few users in a VLC cell. Additionally, since the channel remains fairly constant for VLC systems, channel estimation becomes simpler. NOMA has the ability to handle load distribution and interference cancellation by leveraging channel status information. To further improve its spectrum efficiency, NOMA-based VLC systems can use high-level modulation techniques, such as quadrature amplitude modulation (QAM).

When using NOMA with higher cognitive modulation, there is a chance for increased inter-user interference. Therefore, it is necessary to exercise caution and implement power allocation (PA) strategy accordingly.

In a recent study [54], researchers determined the potential constraint for using higher-order modulation in NOMA-enabled VLC systems. However, existing research primarily focuses on lower-order techniques with an assumption of perfect SIC. Hence, there is a crucial need to address the challenge of handling errors in NOMA-aided VLC systems that use higher-order modulation.

The literature discusses how NOMA approaches can increase the throughput and reliability of VLC systems. Unlike OMA, which limits usage to one user per frequency/timeresource block, NOMA allows multiple clients/users to share those resources, sacrificing a bit of IUI but making more efficient use of available resources [55].

Although VLC has several benefits, there are certain limitations that prevent current technology from fulfilling the needs of 6G networks. The range of visible light signals is limited, which makes the propagation losses significant and weakens the VLC channel when there is a gap between transmitting and receiving devices. Additionally, obstacles can block out emission spectra easily [56].

Moreover, unlike the typical RF counterpart, the VLC channel is not uniform in all directions. Therefore, the shape and position of both transmitting and receiving devices have a significant effect on the channel gains as stated in reference [57].

The quality of the VLC channel varies, affecting the efficiency of advanced MA techniques like NOMA when used with VLC systems. A visual representation of downlink NOMA-VLC for two clients/users is shown in Figure 6.19.

As shown in Figure 6.19, the basic concept of VLC-NOMA is same as in NOMA but the difference is in the communication channel [58]. In VLC-NOMA a visible light source (LED) is used in place of transmitting antenna and visible light detector is used in place of a receiver antenna [59].

Regarding the enhancement of the spectrum efficiency of indoor VLC systems, PD-NOMA can be considered a viable MA approach, as it provides ample bandwidth and allows for notable improvements in throughput and other key performance metrics. With this in mind, NOMA could be viewed as a powerful and effective strategy for MA in VLC systems. A small number of clients/users may be accommodated

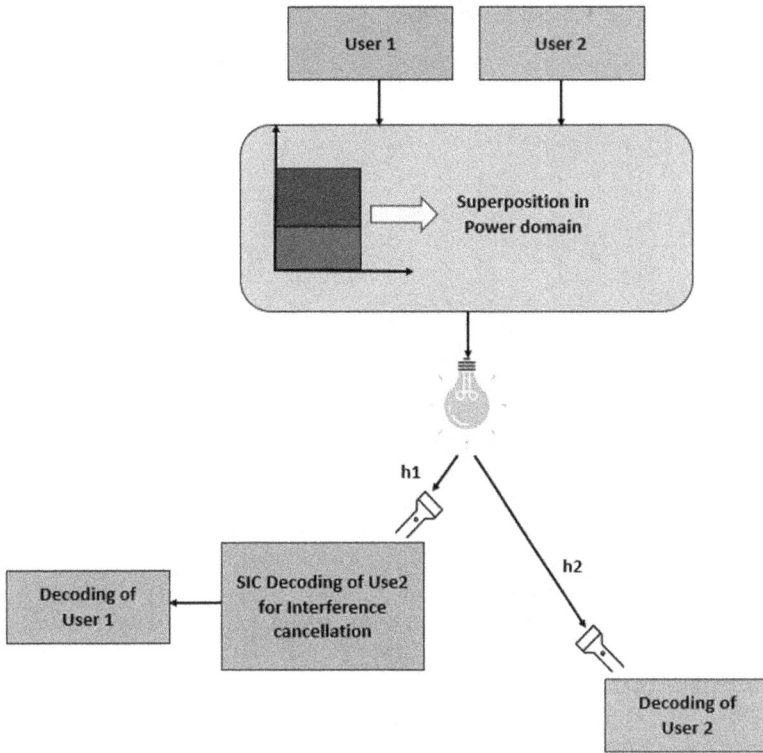

FIGURE 6.19   Downlink NOMA-VLC overview.

using NOMA technology, and this is also true of VLC systems, which employ LEDs as transmitters. For a limited number of clients/users, LED sources serve as entry points in confined environments. In order to effectively manage user power allocation, demultiplexing, and decoding orders via the SIC process, NOMA requires comprehension of channel status information at both the transmitter and receiver sides. The VLC systems, in contrast, enjoy channels that are reasonably stable and don't require as much regular monitoring. Since the LED and PD are so close to one another, NOMA technology excels in VLC networks' high SNR environments.

By adjusting the transmission angles of LEDs and the field-of-view of PDs, differences in channel gains between clients/users in VLC systems may be minimized. NOMA is regarded as the optimal MA strategy in VLC systems to achieve low latency, particularly when user channels exhibit large gain disparities, according to reference [60]. A conceptual design of a NOMA-VLC system is shown in Figure 6.20(b), whereas Figure 6.20(a) shows a simple VLC system [61]. Data are transmitted from a transmitter (an LED) to two receivers in the NOMA-VLC system (Figure 6.20(b)). User 1 experiences greater signal intensity when they are far apart from User 2, while a shorter line of sight to the LED results in weaker signals. Using NOMA techniques, User 2 is given a more powerful h2 to directly decode

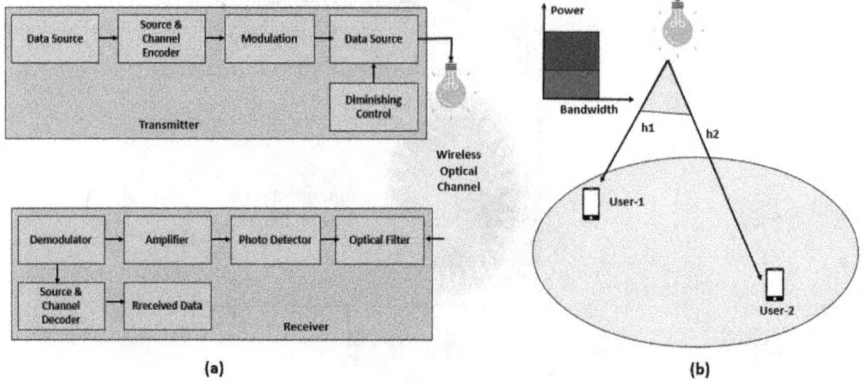

FIGURE 6.20   (a) An overview of a basic VLC system. (b) Conceptual diagram of a NOMA-VLC system.

its message signal. In contrast, h1 receives lower power and must go through SIC phenomenon to decode its required signal [62]. Without a doubt, NOMA schemes outperform the OMA schemes now in use for VLC systems in terms of performance. However, it is impractical to apply NOMA methods to every user since doing so would make SIC computations more difficult, especially as the user base grows. User pairing is used as an alternate solution to this problem, which lowers the difficulty of SIC decoding.

As stated in [63], using a suitable power allocation and user pairing can assist NOMA-VLC systems to work better. Researchers are becoming more and more interested in investigating the viability of NOMA with VLC since it has the potential to replace other options as the default in future cellular networks. The benefits of using cooperative and hybrid NOMA approaches for VLC systems have been emphasized.

Cooperative NOMA schemes can be advantageous for NOMA-VLC systems. The strong user may, however, not choose to use their power to send signals to the weak user. Therefore, one potential answer is to look at ways for the strong user to harvest energy from optical sources and then utilize it to convey messages to the weak user.

If the number of APs used by a VLC system to send power and information to several clients/users is less than the number of clients/users, a serious problem may arise. In this case, choose between OMA and NOMA, for user scheduling and energy harvesting becomes critical. The VLC-NOMA system can meet the required data rates for all clients/users while using less transmitting power since NOMA delivers greater data rates compared to OMA.

In order to utilize redundant data in NOMA systems and balance out poor user-facing co-channel interference, cooperative NOMA is a technique used in RF and VLC networks. It may be used in relay and VLC systems. Clients/users in a two-user situation with a VLC connection can all access data from the RF or VLC network simultaneously. The signal from the stronger user can be decoded and transmitted to the intended receiver through Bluetooth or wi-fi. The weaker user may then combine the signals by employing particular techniques that integrate both RF and VLC signals.

Multiple studies have explored ways to modify and encode signals for RF-NOMA systems in the literature. However, it's worth noting that NOMA-VLC utilizes different modulation and coding schemes, which require novel approaches for successful implementation of these systems.

VLC systems can be unpredictable. Some clients/users may experience poor quality of service due to various factors such as handover overhead, interference between cells, and blockage of line-of-sight. In contrast, others may enjoy good service levels. The solution is a cooperative NOMA-based scheme whereby strong signal clients/users support their weak signal counterparts via RF links. By doing so, a hybrid system can be established that provides consistently reliable service for all clients/users.

Emerging technologies, such as MISO/MIMO techniques, underwater applications, NOMA-based hybrid RF/VLC, intelligent reflecting surfaces (IRSs), unmanned aerial vehicle (UAV), OFDM, and ML techniques are incorporated into NOMA-VLC systems. However, adopting NOMA in VLC systems may result in additional limitations and difficulties that call for appropriate solutions.

Although the majority of the research on NOMA-aided VLC systems focuses on outage probability and sum rate assessments. BER performance may also be taken into account. NOMA-based VLC systems offer several advantages and disadvantages as given below.

Advantages of NOMA-based VLC system

1. **Spectral efficiency:** NOMA allows many clients/users to share the same frequency band by allocating different power levels and superposing their signals. This increases the SE of the system, enabling more clients/users to be accommodated within a given bandwidth.
2. **Increased capacity:** By using NOMA, VLC systems can support a larger number of clients/users compared to traditional OMA schemes. This capacity enhancement is especially beneficial in scenarios with a high density of clients/users or in areas with limited available spectrum.
3. **Fairness:** NOMA enables a flexible allocation of power levels to different clients/users, allowing a trade-off between clients/users with varying channel conditions. This leads to improved fairness in resource allocation, ensuring that all clients/users receive a reasonable QoS.
4. **Lower latency:** NOMA-based VLC systems can achieve lower latency compared to OMA schemes. Since NOMA allows simultaneous transmission, there is no need for time-division multiplexing, resulting in reduced latency and faster data transmission.

Disadvantages of NOMA-based VLC system

1. **Complexity:** Implementing NOMA in VLC systems introduces additional complexity in the receiver design. Decoding multiple overlapping signals requires sophisticated signal processing techniques and receiver

algorithms. This complexity can increase the cost and energy consumption of the system.

2. **Interference:** NOMA relies on superposing signals, which can lead to interference among clients/users. The interference can degrade the overall system performance and require advanced interference cancellation techniques to mitigate its effects.

3. **Channel variability:** NOMA performance heavily depends on the channel conditions of the clients/users. Unequal channel conditions among clients/users can lead to imbalances in the power allocation, affecting the system's overall performance and fairness.

4. **Limited coverage range:** VLC systems operate using visible light, which has a limited coverage range compared to RF-based communication systems. This limitation restricts the coverage area of NOMA-based VLC systems and makes them more suitable for indoor or short-range applications.

It's important to note that the advantages and disadvantages of NOMA-based VLC systems may vary depending on the specific implementation, environmental factors, and the requirements of the application.

6.11 SUMMARY

In this chapter, fundamentals of popular MA schemes have been discussed. Applications, key challenges, opportunities, advantages, and disadvantages also discussed. This chapter covered the latest communication technologies including brief theoretical analysis with integration of NOMA. The chapter also highlights the new limits and obstacles in implementing NOMA in VLC systems. This chapter give a road map of various MA schemes form 1G to 5G and beyond that is very helpful in gaining a basic conceptual knowledge of various MA schemes used in 5G and beyond communication technologies.

REFERENCES

1. Rappaport, T. S. Wireless Communications: Principles and Practice. Pearson Publication, Second Edition, 2010.
2. Shah, A.F., Qasim, A.N., Karabulut, M.A., Ilhan, H., & Islam, M. (2021). Survey and Performance Evaluation of Multiple Access Schemes for Next-Generation Wireless Communication Systems. *IEEE Access*, *9*, 113428–113442.
3. Chege, S., Walingo, T., & Derraz, F. (2023). Codebook Design for PD-SCMA NOMA Systems. *IEEE Access*, *11*, 13421–13431.
4. Zheng, Y., Hou, X., Wang, H., Jiang, M., & Zhang, S. (2023). A Knowledge-Based Deep Learning Detection Scheme for Downlink SCMA Systems. *IEEE Wireless Communications Letters*, *12*, 486–490.
5. de Oliveira Torres, F., de Santiago Júnior, V.A., da Costa, D.B., Cardoso, D.L., & de Oliveira, R.C. (2023). Throughput Maximization for a Multicarrier Cell-Less NOMA Network: A Framework Based on Ensemble Metaheuristics. *IEEE Transactions on Wireless Communications*, *22*, 348–361.

6. Rashid, B., Ahmad, A., Ali, S., Kaleem, Z., Alkhayyat, A., & Yuen, C. (2023). Energy Efficient Resource Allocation for Uplink MC-NOMA Based Heterogeneous Small Cell Networks with Wireless Backhaul. *IEEE Transactions on Vehicular Technology, 72*, 3419–3429.

7. Andrews, J.G., & Meng, T.H. (2003). Optimum Power Control for Successive Interference Cancellation with Imperfect Channel Estimation. *IEEE Trans. Wirel. Commun., 2*, 375–383.

8. R. C. Kizilirmak, 'Non-Orthogonal Multiple Access (NOMA) for 5G Networks', Towards 5G Wireless Networks - A Physical Layer Perspective. *InTech*, Dec. 14, 2016. doi: 10.5772/66048.

9. Shukla, A., Kumar, M., & Gupta, M. (2023). Intelligent Reflecting Surfaces Assisted Full-Duplex IDMA Communication Network. *Wireless Personal Communications, 130*, 469–479.

10. Millar, G., Kulhandjian, M., Alaca, A., Alaca, S., D'amours, C., & Yanikomeroglu, H. (2022). Low-Density Spreading Design Based on an Algebraic Scheme for NOMA Systems. *IEEE Wireless Communications Letters, 11*, 698–702.

11. Tang, W., Wang, L., Wang, Z., Han, G., Zhao, J., & Yue, X. (2022). Research on Improved Multiple Users' Detection Algorithm in Pattern Division Multiple Access System. *Digit. Signal Process., 133*, 103877.

12. Saito, Y., Kishiyama, Y., Benjebbour, A., Nakamura, T., Li, A., & Higuchi, K. (2013). Non-Orthogonal Multiple Access (NOMA) for Cellular Future Radio Access. *2013 IEEE 77th Vehicular Technology Conference (VTC Spring)*, 1–5.

13. Aldababsa, M., Toka, M., Gökceli, S., Karabulut-Kurt, G., & Kucur, O. (2018). A Tutorial on Nonorthogonal Multiple Access for 5G and Beyond. *Wireless Communication and Mobile Computing, 2018*, 9713450.

14. Alkhawatrah, M. (2023). The Performance of Supervised Machine Learning Based Relay Selection in Cooperative NOMA. *IEEE Access, 11*, 1570–1577.

15. Ding, Z., Peng, M., & Poor, H.V. (2014). Cooperative Non-Orthogonal Multiple Access in 5G Systems. *IEEE Communications Letters, 19*, 1462–1465.

16. Li, L., Wang, L., & Hanzo, L. (2012). Differential Interference Suppression Aided Three-Stage Concatenated Successive Relaying. *IEEE Transactions on Communications, 60*, 2146–2155.

17. Huang, J., Peng, K., Pan, C., Yang, F., & Jin, H. (2014). Scalable Video Broadcasting Using Bit Division Multiplexing. *IEEE Transactions on Broadcasting, 60*, 701–706.

18. Dai, L., Wang, B., Ding, Z., Wang, Z., Chen, S., & Hanzo, L.H. (2018). A Survey of Non-Orthogonal Multiple Access for 5G. *IEEE Communications Surveys & Tutorials, 20*, 2294–2323.

19. Hayat, O., Ngah, R., & Hashim, S.Z. (2020). Multi-user Shared Access (MUSA) Procedure for Device Discovery in D2D Communication. *Telecommunication Systems, 76*, 291–297.

20. Yuan, Z., Yu, G., Li, W., Yuan, Y., Wang, X., & Xu, J. (2016). Multi-User Shared Access for Internet of Things. *2016 IEEE 83rd Vehicular Technology Conference (VTC Spring)*, 1–5.

21. Eid, E.M., Fouda, M.M., Eldien, A.S., & Tantawy, M.M. (2017). Performance analysis of MUSA with different spreading codes using ordered SIC methods. *2017 12th International Conference on Computer Engineering and Systems (ICCES)*, 101–106.

22. Gui, G., Liu, M., Kato, N., Adachi, F., & Tang, F. (2020). 6G: Opening New Horizons for Integration of Comfort, Security, and Intelligence. *IEEE Wireless Communications, 27*, 126–132.

23. Mao, Y., Clerckx, B., & Li, V.O. (2017). Rate-splitting Multiple Access for Downlink Communication Systems: Bridging, Generalizing, and Outperforming SDMA and NOMA. *Eurasip Journal on Wireless Communications and Networking, 2018*.

24. Mao, Y., Dizdar, O., Clerckx, B., Schober, R., Popovski, P., & Poor, H.V. (2022). Rate-Splitting Multiple Access: Fundamentals, Survey, and Future Research Trends. *IEEE Communications Surveys & Tutorials, 24*, 2073–2126.

25. Mishra, A., Mao, Y., Dizdar, O., & Clerckx, B. (2022). Rate-Splitting Multiple Access for 6G—Part I: Principles, Applications and Future Works. *IEEE Communications Letters, 26*, 2232–2236.

26. Xu, J., Dizdar, O., & Clerckx, B. (2023). Rate-Splitting Multiple Access for Short-Packet Uplink Communications: A Finite Blocklength Analysis. *IEEE Communications Letters, 27*, 517–521.

27. Park, J., Choi, J., Lee, N., Shin, W., & Poor, H.V. (2021). Rate-Splitting Multiple Access for Downlink MIMO: A Generalized Power Iteration Approach. *IEEE Transactions on Wireless Communications, 22*, 1588–1603.

28. Lee, B., & Shin, W. (2023). Max-Min Fairness Precoder Design for Rate-Splitting Multiple Access: Impact of Imperfect Channel Knowledge. *IEEE Transactions on Vehicular Technology, 72*, 1355–1359.

29. Wang, Y., Wong, V.W., & Wang, J. (2023). Flexible Rate-Splitting Multiple Access With Finite Blocklength. *IEEE Journal on Selected Areas in Communications, 41*, 1398–1412.

30. Wu, M., Gao, Z., Huang, Y., Xiao, Z., Ng, D.W., & Zhang, Z. (2022). Deep Learning-Based Rate-Splitting Multiple Access for Reconfigurable Intelligent Surface-Aided Tera-Hertz Massive MIMO. *IEEE Journal on Selected Areas in Communications, 41*, 1431–1451.

31. Wang, P., Li, Y., Lin, Z., Shirvanimoghaddam, M., Park, O., Park, G., & Vucetic, B. (2023). Analysis of Rateless Multiple Access Scheme with Maximum Likelihood Decoding in an AWGN Channel. *IEEE Transactions on Wireless Communications*.

32. S. Han, C.-L. I, Z. Xu, and Q. Sun, (2014). Energy efficiency and Spectrum Efficiency Co-design: From NOMA to Network NOMA. *IEEE MMTC E-Letter, 9*, 5, 21–24.

33. Tabassum, H., Hossain, E., & Hossain, J. (2016). Modeling and Analysis of Uplink Non-Orthogonal Multiple Access in Large-Scale Cellular Networks Using Poisson Cluster Processes. *IEEE Transactions on Communications, 65*, 3555–3570.

34. Shin, W., Vaezi, M., Lee, B., Love, D.J., Lee, J., & Poor, H.V. (2016). Non-Orthogonal Multiple Access in Multi-Cell Networks: Theory, Performance, and Practical Challenges. *IEEE Communications Magazine, 55*, 176–183.

35. Rezvani, S., Yamchi, N.M., Javan, M.R., & Jorswieck, E.A. (2020). Resource Allocation in Virtualized CoMP-NOMA HetNets: Multi-Connectivity for Joint Transmission. *IEEE Transactions on Communications, 69*, 4172–4185.

36. R. Hadani and S. S. Rakib. (2018). Multiple access in an orthogonal time frequency space communication system. U.S. Patent 9,967,758.

37. Monk, A.M., Hadani, R., Tsatsanis, M.K., & Rakib, S. (2016). OTFS - Orthogonal Time Frequency Space. *ArXiv, abs/1608.02993*.

38. Hadani, R., Rakib, S., Tsatsanis, M.K., Monk, A.M., Goldsmith, A.J., Molisch, A.F., & Calderbank, A.R. (2017). Orthogonal Time Frequency Space Modulation. *2017 IEEE Wireless Communications and Networking Conference (WCNC)*, 1–6.

39. Raviteja, P., Hong, Y., Viterbo, E., & Biglieri, E. (2019). Practical Pulse-Shaping Waveforms for Reduced-Cyclic-Prefix OTFS. *IEEE Transactions on Vehicular Technology, 68*, 957–961.

40. Surabhi, G.D., Augustine, R.M., & Chockalingam, A. (2018). On the Diversity of Uncoded OTFS Modulation in Doubly-Dispersive Channels. *IEEE Transactions on Wireless Communications, 18*, 3049–3063.

41. Khammammetti, V., & Mohammed, S.K. (2019). OTFS-Based Multiple-Access in High Doppler and Delay Spread Wireless Channels. *IEEE Wireless Communications Letters, 8*, 528–531.

42. Le, M.T., Ferrante, G.C., Quek, T.Q., & Di Benedetto, M. (2017). Fundamental Limits of Low-Density Spreading NOMA With Fading. *IEEE Transactions on Wireless Communications*, *17*, 4648–4659.
43. Tse D. & Viswanath, P. (2005). Fundamentals of Wireless Communication. Cambridge, U.K.: Cambridge University Press.
44. Cover T. M., & Thomas, J. A. (2012). Elements of Information Theory. Hoboken, NJ, USA: Wiley.
45. Ding, Z., Schober, R., Fan, P., & Vincent Poor, H. (2019). OTFS-NOMA: An Efficient Approach for Exploiting Heterogenous User Mobility Profiles. *IEEE Transactions on Communications*, *67*, 7950–7965.
46. Ding, Z. (2019). Robust Beamforming Design for OTFS-NOMA. *IEEE Open Journal of the Communications Society*, *1*, 33–40.
47. Chatterjee, A., Rangamgari, V., Tiwari, S., & Das, S.S. (2020). Nonorthogonal Multiple Access with Orthogonal Time–Frequency Space Signal Transmission. *IEEE Systems Journal*, *15*, 383–394.
48. Wang, P., Xiao, J., & Ping, L. (2006). Comparison of Orthogonal and Non-orthogonal Approaches to Future Wireless Cellular Systems. *IEEE Vehicular Technology Magazine*, *1*, 4–11.
49. Al-Eryani, Y.F., & Hossain, E. (2019). The D-OMA Method for Massive Multiple Access in 6G: Performance, Security, and Challenges. *IEEE Vehicular Technology Magazine*, *14*, 92–99.
50. Hashemi, R., Beyranvand, H., Mili, M.R., Khalili, A., Tabassum, H., & Ng, D.W. (2021). Energy Efficiency Maximization in the Uplink Delta-OMA Networks. *IEEE Transactions on Vehicular Technology*, *70*, 9566–9571.
51. Feng, L., Hu, R.Q., Wang, J., Xu, P., & Qian, Y. (2016). Applying VLC in 5G Networks: Architectures and Key Technologies. *IEEE Network*, *30*, 77–83.
52. Marshoud, H., Kapinas, V.M., Karagiannidis, G.K., & Muhaidat, S.H. (2015). Non-Orthogonal Multiple Access for Visible Light Communications. *IEEE Photonics Technology Letters*, *28*, 51–54.
53. Shi, J., He, J., Wu, K., & Ma, J. (2019). Enhanced Performance of Asynchronous Multi-Cell VLC System Using OQAM/OFDM-NOMA. *Journal of Lightwave Technology*, *37*, 5212–5220.
54. Almohimmah, E.M., & Alresheedi, M.T. (2020). Error Analysis of NOMA-Based VLC Systems with Higher Order Modulation Schemes. *IEEE Access*, *8*, 2792–2803.
55. Arfaoui, M.A., Ghrayeb, A., Assi, C.M., & Qaraqe, M.K. (2021). CoMP-Assisted NOMA and Cooperative NOMA in Indoor VLC Cellular Systems. *IEEE Transactions on Communications*, *70*, 6020–6034.
56. Arfaoui, M.A., Soltani, M.D., Tavakkolnia, I., Ghrayeb, A., Assi, C.M., Safari, M., & Haas, H. (2020). Measurements-Based Channel Models for Indoor LiFi Systems. *IEEE Transactions on Wireless Communications*, *20*, 827–842.
57. Soltani, M.D., Arfaoui, M.A., Tavakkolnia, I., Ghrayeb, A., Safari, M., Assi, C.M., Hasna, M.O., & Haas, H. (2018). Bidirectional Optical Spatial Modulation for Mobile Users: Toward a Practical Design for LiFi Systems. *IEEE Journal on Selected Areas in Communications*, *37*, 2069–2086.
58. Mohsan, S.A., Sadiq, M., Li, Y., Shvetsov, A.V., Shvetsova, S.V., & Shafiq, M. (2023). NOMA-Based VLC Systems: A Comprehensive Review. *Sensors (Basel, Switzerland)*, *23*.
59. Marshoud, H., Sofotasios, P.C., Muhaidat, S.H., & Karagiannidis, G.K. (2016). Multi-user techniques in visible light communications: A survey. *2016 International Conference on Advanced Communication Systems and Information Security (ACOSIS)*, 1–6.

60. Yin, L., Popoola, W.O., Wu, X., & Haas, H. (2016). Performance Evaluation of Non-Orthogonal Multiple Access in Visible Light Communication. *IEEE Transactions on Communications, 64*, 5162–5175.

61. Yang, Z., Xu, W., & Li, Y. (2017). Fair Non-Orthogonal Multiple Access for Visible Light Communication Downlinks. *IEEE Wireless Communications Letters, 6*, 66–69.

62. Dogra, T., & Bharti, M.R. (2022). User Pairing and Power Allocation Strategies for Downlink NOMA-based VLC Systems: An Overview. *AEU - International Journal of Electronics and Communications, 149*, 154184.

63. Obeed, M., Salhab, A.M., Alouini, M., & Zummo, S.A. (2018). On Optimizing VLC Networks for Downlink Multi-User Transmission: A Survey. *IEEE Communications Surveys & Tutorials, 21*, 2947–2976.

7 Trends in Intelligent Communication and Wireless Technologies

*Aman Jolly, Vikas Pandey, and
Kaibalya Prasad Bhuyan*

7.1 INTRODUCTION

Intelligent communication and wireless technologies refer to the combination of various technologies that are used to transmit information, data, and voice wirelessly, with the help of artificial intelligence (AI) and machine learning (ML) algorithms. These technologies enable the communication systems to become intelligent, allowing for faster and more accurate decision-making, real-time processing, and adaptability to changing conditions.

Here is a brief overview of the emerging trends in these technologies:

1. 5G technology: 5G is the latest generation of wireless technology that promises to deliver faster, more reliable, and more secure wireless communication. It offers improved network efficiency and reduced latency compared to previous generations.
2. Internet of Things (IoT): IoT refers to the interconnected network of physical devices, vehicles, home appliances, and other items embedded with electronics, software, sensors, and network connectivity, enabling these objects to collect and exchange data.
3. AI: AI is the development of computer systems that can perform tasks that would normally require human intervention, such as speech recognition, decision-making, and pattern recognition. AI is being increasingly integrated into wireless communication systems to enhance the performance and efficiency of these systems.
4. Edge computing: Edge computing refers to the processing of data and computation at the edge of the network, closer to the source of the data, rather than in a centralized data center or cloud. This allows for faster and more efficient processing of data, reducing the load on centralized data centers.

The major contributions in this chapter are as follows:

- Analyze various trends in intelligent communication and wireless technologies.
- Perform the literature review on current trends in intelligent communication and wireless technologies (Figure 7.1).

DOI: 10.1201/9781003407836-7

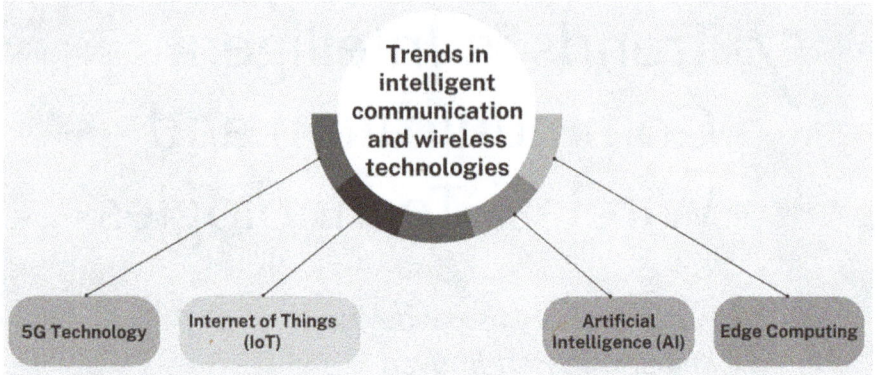

FIGURE 7.1   Trends in intelligent communication and wireless technologies.

7.2   EMERGING TRENDS

Emerging trends in intelligent communication and wireless technologies have the potential to revolutionize the way we communicate and interact with the world. These technologies, such as 5G, IoT, AI, Edge Computing, and enhanced Mobile Broadband (eMBB) have the potential to bring about significant advancements in fields such as medicine, transportation, and entertainment. Figure 7.2 shows the cumulative trend of AI and 5G gathered by country from Google Trends between 1/1/2014 to 7/2/1023. This also indicates the correlation between the Trends and countries on the different trends for the development of new policies. However, it is also important to be mindful of the limitations and challenges associated with these technologies, such as security and bias, and to take steps to address these challenges. Ultimately, the impact of these emerging trends will depend on how they are designed, developed, and implemented, and it will be important to ensure that they

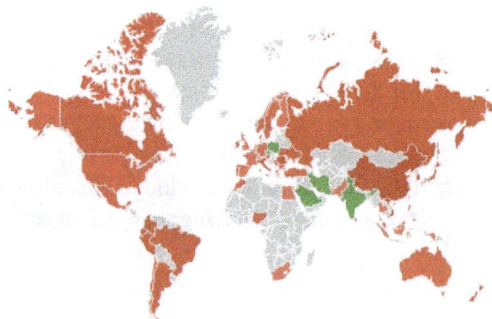

FIGURE 7.2   Worldwide trends by country between 01/01/2004–07/02/2023 from Google Trends.

are used in a way that benefits society as a whole. In this section we would be reviewing them on various aspects

7.2.1   5G TECHNOLOGY

With the increasing demand for higher bandwidth and lower latency in mobile networks, 5G technology has seen rapid growth in recent years. 5G networks are expected to support a wide range of new applications and services, from enhanced mobile broadband to IoT and industrial automation. 5G technology can revolutionize the present system by providing high-speed, low-latency connectivity that enables new applications and services. One example of a use case is the development of smart cities, where 5G networks can be used to support connected transportation systems, smart energy systems, and other infrastructure-level services. Additionally, 5G can also enable advancements in fields such as virtual reality and augmented reality, making it possible to create new forms of entertainment and communication.

The 5G network is revolutionizing the healthcare industry with its unparalleled speed and low latency. A proof-of-concept study was carried out in two phases, with the first phase focusing on examining patients using portable ultrasound equipment in various locations and the second phase focusing on remote examination with health professionals in different states. Oliveira et al. (2023) evaluated the connectivity and capacity of the 5G private network to transmit large amounts of data from remote exams [1]. To ensure financial sustainability, Wu et al. (2023) analyzed the profitability mechanism of live sporting event streaming (LSES) based on stimulus-organism-response (SOR) theory and two-sided market characteristics [2]. Researchers have also explored the global research status and trends in microperimetry (Ming et al., 2023) and provided a comprehensive overview of 5G handover management (Wang et al., 2023) [3, 4]. In addition, a bibliometric study was performed using CiteSpace and VOSviewer (Du et al., 2023), and the application of AI technology in optical communication networks and 5G was discussed (Zhang et al., 2023) [6, 5]. Sathyanarayana et al. (2023) proposed a unique 5G-LAN orchestrator (5GLoR) for the first time [7]. Other influential works include Fan et al. (2023), Lin (2023), and Doddamane et al. (2023) [8–10]. The potential of 5G in transforming healthcare is truly remarkable and holds great promise for improving patients' quality of life.

There have been several recent research developments in the area of 5G technology. Some of the most notable include

1. Integration with existing infrastructure: There has been significant research into the integration of 5G networks with existing infrastructure, such as 4G networks and Wi-Fi, in order to provide a more seamless and efficient user experience. This has included the development of new technologies, such as dual-connectivity and network slicing, which allow for the efficient use of multiple networks in a single device.
2. Development of new 5G standards: There has been ongoing research into the development of new 5G standards, with a focus on areas such as low

latency, high-speed connectivity, and support for new use cases, such as autonomous vehicles and industrial IoT.

3. Security and privacy: With the increasing use of 5G networks, there has been growing concern about the security and privacy implications of these networks. As a result, there has been significant research into the development of more secure and private 5G networks, including the use of encryption, secure communications protocols, and secure device management.

4. 5G for Industrial IoT: There has been a significant amount of research into the use of 5G technology for industrial IoT, with a focus on areas such as smart manufacturing, predictive maintenance, and real-time control.

5. 5G for autonomous vehicles: There has been a significant amount of research into the use of 5G technology for autonomous vehicles, with a focus on areas such as vehicle-to-vehicle communications, real-time mapping, and high-definition video streaming.

One of the research directions in 5G technology is the development of more efficient and secure networks, as 5G networks are expected to support a large number of devices and applications. Additionally, there is a need for further research on the deployment and integration of 5G networks with existing infrastructure, including 4G networks and Wi-Fi. Another research direction is in the area of developing new use cases for 5G technology, as well as improving the performance and reliability of 5G networks in challenging environments, such as in rural areas or in high-density urban environments.

7.2.2 INTERNET OF THINGS (IoT)

IoT has the potential to revolutionize the present system by enabling the development of smart homes, smart cities, and connected industrial systems. For example, IoT devices can be used to monitor and control energy consumption in homes and buildings, and to manage traffic flow in cities, helping to optimize energy usage and reduce emissions. Additionally, the IoT can be used to track the location and status of assets in supply chains, making it possible to improve the efficiency and transparency of these systems.

There has been a significant amount of recent research in the field of IoT. Some of the most notable areas of research include

1. Edge computing: There has been a growing interest in edge computing for IoT, with a focus on bringing computation and storage closer to the source of data, in order to reduce latency and increase efficiency.

2. Security and privacy: With the increasing deployment of IoT devices, there has been growing concern about the security and privacy implications of these devices. As a result, there has been significant research into the development of more secure and private IoT systems, including the use of encryption, secure communications protocols, and secure device management.

3. Low-power, wide-area networks (LPWANs): There has been a growing interest in LPWANs for IoT, with a focus on providing low-power,

    long-range connectivity for IoT devices, particularly in rural and remote areas.

4. AI and ML: There has been a significant amount of research into the use of AI and ML for IoT, with a focus on areas such as predictive maintenance, real-time control, and improved decision-making.

5. IoT for healthcare: There has been a growing interest in the use of IoT for healthcare, with a focus on areas such as remote patient monitoring, telemedicine, and wearable medical devices. The IoT has the potential to revolutionize the medical industry by enabling communication among objects to collect and analyze data to make informed decisions that can improve the quality of life. In order to achieve this, researchers have been developing an intelligent framework for the Internet of Medical Things (IoMT) that incorporates advanced technologies such as ensemble learning, genetic algorithms, blockchain, and various U-Net based architectures.

These studies aim to focus on the usability aspect of the system and apply it to specific scenarios, such as the early detection of sepsis using electronic health records.

The IoT enables communication among objects to collect information and make decisions to improve the quality of life [11]. In the quest of training complicated medical data for IoMT scenarios Belhadi et al. (2023) develop an end-to-end intelligent framework that incorporates ensemble learning, genetic algorithms, blockchain technology, and various U-Net based architectures [12]. To achieve such a system, Alam et al. (2023) focus on several aspects, most notably the usability aspect of deploying it using low-end devices [13]. They introduce one such application, namely, for the early detection of sepsis using electronic health records [13]. Geng et al. (2023) present an overview of the existing works on human decision-making and human machine collaboration within the scope of signal processing and information fusion [14]. Tselebis et al. (2023) assess in Google Trends searches for 21 precocious puberty (PP)-related terms in English internationally (which practically dwarf searches in other languages), in the years 2017–2021 [15].

In addition, there are research works on human decision-making and human-machine collaboration, as well as a framework for a diabetes prediction scenario. The proposed IoMT framework is presented with a diabetes prediction scenario [16].

The use of blockchain and AI in combination with IoT can also help to address the current challenges in maintaining the integrity, transparency, reliability, and trustworthiness of operations for small and medium-sized enterprises (SMEs). Security frameworks have also been proposed to address the security concerns related to the use of IoT in the medical field. Till now, it is hard to maintain operations of SMEs' integrity, transparency, reliability, provenance, availability, and trustworthiness between two different enterprises due to the current nature of centralized server-based infrastructure. Khan et al. (2023) propose a novel and secure framework with a standardized process hierarchy/lifecycle for distributed SMEs using collaborative techniques of blockchain, IoT, and AI with ML [17]. Ghubaish et al. (2023) propose a security framework that combines several security techniques [18]. Other influential work includes Kang et al. (2023) [19, 20].

One of the research gaps in IoT is the development of more secure and reliable networks, as IoT devices are expected to collect and transmit large amounts of sensitive data. The interconnected nature of IoT devices leaves them vulnerable to cyberattacks. Additionally, there is a need for further research on the development of more efficient and effective ways to process and analyze the data generated by IoT devices. Another research gap is in the area of developing new use cases for IoT, such as in the context of smart cities and connected homes, as well as improving the interoperability and compatibility of IoT devices. The compatibility between different IoT devices can be a challenge for users. Example: A smart home system that is incompatible with other devices in the home can limit the overall user experience.

7.2.3 Artificial Intelligence (AI)

AI and ML have the potential to revolutionize the present system by automating processes and making decision-making more efficient and accurate. One example of a use case is the development of healthcare applications, where AI can be used to diagnose diseases and predict patient outcomes. Additionally, AI can be used to automate tasks in industries such as manufacturing and finance, improving efficiency and reducing the risk of errors.

The application of AI in the medical field is a rapidly growing and evolving area of research. Researchers are utilizing cutting-edge bibliometrics software and comprehensive literature surveys to gain a deeper understanding of the latest developments and trends in this field. The results of these studies have the potential to not only provide valuable insights into the state of the industry, but also to have a real-world impact by shaping the future of the smart medical industry.

For example, the analysis of the development trends and spatial non-equilibrium characteristics of the intelligent smart medical industry in the Yangtze River Economic Belt [5] has the potential to significantly inform policy decisions, optimizing the spatial layout of the integrated development of this industry. The examination of AI applications in mental health [21] highlights the wide range of methodologies and techniques being used, from ML and data analytics to natural language processing and decision support, and sheds light on the challenges and open research issues in this field.

Furthermore, the analysis of academic literature on AI in diabetic retinopathy (Poly et al., 2023) and the overview of AI in hand and wrist surgery (Miller et al., 2023) demonstrate the potential of AI to revolutionize medical treatments and improve patient outcomes. The authors provide an update on the current state of AI in these fields and offer suggestions for future directions and applications [22, 23].

The papers in this field analyze the use of AI in various medical applications such as sepsis, diabetic retinopathy, FDA approved AI/ML-enabled medical devices, mental health, and hand and wrist surgery, using bibliometric software, literature surveys, and FDA database analysis. The authors aim to understand the development status, identify core hotspots and future trends, and provide policy implications for optimizing the spatial layout of the smart medical industry. These papers cover a wide range of topics including ML, data analytics, natural language processing, decision support, and explainable AI, and highlight open research issues and chal-lenges [24–28].

In conclusion, the research being conducted in the field of AI in medicine is providing valuable insights into the latest trends and developments, and has the potential to shape the future of the industry. These studies demonstrate the impor-tance of continued investment and research in this field, which holds great promise for improving the health and well-being of people everywhere.

There have been many recent developments in the field of AI and ML. Some of the most notable areas of research include

1. Deep learning: Deep learning, a subfield of ML that is inspired by the struc-ture and function of the human brain, has been a rapidly growing area of research in recent years. The field has seen significant advancements in areas such as computer vision, speech recognition, and natural language processing.
2. Explainable AI (XAI): There has been a growing interest in developing AI systems that can be transparent and explainable, particularly in applications such as healthcare, finance, and criminal justice. Researchers are exploring techniques such as model interpretability, feature attribution, and model distillation to create more XAI models.
3. Reinforcement learning: Reinforcement learning, a subfield of ML that focuses on decision-making problems, has been a rapidly growing area of research in recent years. The field has seen significant advancements in areas such as game playing, autonomous control, and decision-making in uncertain environments.
4. Generative models: Generative models, a subfield of ML that focuses on creating new data from existing data, have been a rapidly growing area of research in recent years. The field has seen significant advancements in areas such as computer vision, speech synthesis, and natural language processing.
5. Transfer learning: Transfer learning, a subfield of ML that focuses on trans-ferring knowledge from one domain to another, has been a rapidly growing area of research in recent years. The field has seen significant advancements in areas such as computer vision, speech recognition, and natural language processing.

One of the research gaps in AI and ML is the development of more accurate and reliable algorithms, as well as addressing the potential biases and ethical concerns associated with the use of AI. Additionally, there is a need for further research on the integration of AI and ML with existing systems, such as in the context of healthcare and finance, as well as on the development of new use cases for AI and ML. The algorithms used for these technologies are only as good as the data they are trained on. This can lead to biased outcomes if the data used for training is biased. Another limitation is the lack of transparency in AI and ML decision-making, which can make it difficult to understand why certain decisions are made. Example: An AI system that is trained on data that includes racial bias can result in racially biased outcomes when making decisions.

7.2.4 EDGE COMPUTING

Edge computing can revolutionize the present system by bringing computing closer to the source of data, reducing latency and improving the performance of applications. One example of a use case is in the industrial sector, where edge computing can be used to process data from IoT devices in real time, allowing for faster decision-making and improved control over industrial processes. Additionally, edge computing can be used in the context of autonomous vehicles, allowing for real-time processing of data from sensors and cameras.

There has been a significant amount of recent research in the field of edge computing. Some of the most notable areas of research include

1. Resource optimization: Researchers are exploring techniques for optimizing resource utilization in edge computing systems, such as algorithms for task scheduling, data placement, and network management.
2. Security and privacy: With the increasing deployment of edge computing systems, there has been growing concern about the security and privacy implications of these systems. As a result, there has been significant research into the development of more secure and private edge computing systems, including the use of encryption, secure communications protocols, and secure device management.
3. Energy efficiency: There has been a growing interest in making edge computing systems more energy efficient, particularly in the context of IoT and mobile devices. Researchers are exploring techniques such as energy-aware task scheduling, energy-efficient data processing, and power management algorithms.
4. Network and communication: There has been a growing interest in improving the performance and reliability of edge computing systems, particularly in terms of network and communication. Researchers are exploring techniques such as network virtualization, software-defined networking (SDN), and wireless mesh networks.
5. Interoperability: There has been a growing interest in making edge computing systems more interoperable, so that data can be seamlessly transferred between different systems and devices. Researchers are exploring techniques such as data exchange protocols, data standardization, and data integration.

Various technological project related to the edge computing and 5G network was explored by many researchers. Oztoprak et al. (2023) discuss the challenges in the telecommunication field from both academic and industry perspectives, focusing on the transformation toward edge-cloud continuum and the evolution of cloud-native computing, software-defined networking (SDN), and network automation platforms [29]. Liu et al. (2023) provide a comprehensive summary of recent advances in aqueous Cs removal through a bibliometric analysis [30]. Castro et al. (2023) created a taxonomy of microservices on edge computing, exploring their architecture approaches, features, composition, and offloading modes [31]. Ming et al. (2023) explore the global research status and trends in microperimetry [32]. Tian et al. (2023) propose a generic neuromorphic framework for edge healthcare and biomedical applications, evaluated on various tasks including EEG-based epileptic seizure prediction, ECG-based arrhythmia detection, and EMG-based hand gesture recognition [33]. Li et al. (2023) propose an orchestration framework, for application tasks on an edge system that involves commercial and personal edge devices [31]. Ming (2023) propose a vehicle-intelligent control system based on edge computing and deep learning [32]. Cornblath et al. (2023) studied 11 patients who underwent cortico-cortical evoked potential (CCEP) mapping for surgical evaluation of drug-resistant epilepsy [34]. Wu et al. (2023) conducted a comprehensive survey on the existing work of optimized federated learning (FL) models, frameworks, and algorithms with a focus on their network topologies. Seisa et al. (2023) is also mentioned as an influential work in [20] and [33].

One of the current limitations of edge computing is the limited processing power and storage capacity of edge devices, which can limit the types of applications that can be run on these devices. Another limitation is the high cost of deploying edge computing infrastructure, which can be a barrier for some organizations. Example: A smart city application that requires significant processing power and storage capacity may not be able to run on edge devices and would need to be processed in the cloud.

7.3 TABULAR ANALYSIS

Sr. No.	Trends	Pros	Cons
1	5G Technology	1. High-speed connectivity: 5G networks provide high-speed connectivity, enabling new applications and services that require high bandwidth and low latency. 2. Improved coverage: 5G networks provide improved coverage, reducing the number of dead spots and improving network availability. 3. Increased capacity: 5G networks provide increased capacity, allowing for more devices and applications to be connected to the network.	1. High implementation costs: The implementation of 5G networks is expensive, requiring significant investment from both operators and governments. 2. Interoperability issues: 5G networks may face interoperability issues with existing 4G networks, requiring careful planning and coordination. 3. Spectrum availability: The availability of spectrum for 5G networks is limited, requiring governments to allocate and manage spectrum carefully.
2	Internet of Things (IoT)	1. Increased connectivity: IoT technologies provide increased connectivity, enabling the connection of various devices and sensors.	1. Security concerns: IoT technologies face significant security concerns, as the connected devices and systems are vulnerable to cyberattacks.

(Continued)

Sr. No.	Trends	Pros	Cons
		2. Real-time data collection: IoT technologies enable real-time data collection, allowing for the monitoring and analysis of various systems and processes. 3. Improved efficiency: IoT technologies can improve efficiency by automating various processes and tasks.	2. Interoperability issues: IoT technologies may face interoperability issues, as different devices and systems may use different communication protocols and technologies. 3. High implementation costs: The implementation of IoT technologies can be expensive, requiring significant investment in devices, sensors, and systems.
3	AI and ML	1. Improved decision-making: AI and ML algorithms can improve decision-making by providing real-time insights and recommendations. 2. Increased efficiency: AI and ML algorithms can increase efficiency by automating various processes and tasks. 3. Personalization: AI and ML algorithms can provide personalized experiences, based on individual preferences and behaviors.	1. Bias and fairness concerns: AI and ML algorithms may introduce biases and unfairness into the decision-making process, requiring careful evaluation and mitigation. 2. Security concerns: AI and ML algorithms may be vulnerable to cyberattacks, requiring secure and robust implementations. 3. Complexity: AI and ML algorithms can be complex and difficult to understand, requiring significant expertise and resources to implement and maintain.
4	Edge Computing	1. Reduced latency: Edge computing provides low-latency processing of data, allowing for faster and more responsive systems. 2. Improved security: Edge computing provides improved security, as the processing of data is done close to the source, reducing the risk of cyber-attacks. 3. Increased scalability: Edge computing provides increased scalability, allowing for the processing of large amounts of data in a distributed manner.	1. High implementation costs: The implementation of edge computing can be expensive, requiring significant investment in hardware and software. 2. Complexity: Edge computing can be complex, requiring significant expertise and resources to implement and maintain. 3. Interoperability issues: Edge computing may face interoperability issues, as different systems and devices may use different technologies and standards.

7.4 USE CASES

There are a few examples where multiple emerging trends in intelligent communication and wireless technologies are being used collectively:

1. Smart healthcare: In this field, 5G networks are being used to provide high-speed connectivity to medical devices, while IoT technologies are being used to connect and gather data from various medical sensors and devices. Additionally, AI and ML algorithms are being used to analyze and make sense of the data gathered from these devices to provide personalized and real-time health recommendations. Edge computing is also being used to process and analyze data close to the source, providing low-latency and secure data processing for critical health applications.
2. Autonomous cars: In the field of autonomous cars, 5G networks are being used to provide high-speed and low-latency connectivity between the car and the surrounding infrastructure, while IoT technologies are being used to connect various sensors and devices within the car. AI and ML algorithms are being used to process and analyze the data from these devices to make real-time driving decisions, and edge computing is being used to provide fast and secure processing of data for critical safety applications.
3. Smart manufacturing: In the field of smart manufacturing, IoT technologies are being used to connect various industrial devices and machines, while 5G networks are being used to provide high-speed connectivity to these devices. AI and ML algorithms are being used to analyze and make sense of the data gathered from these devices to provide real-time production optimization, and edge computing is being used to provide low-latency processing of data for critical industrial control applications.

These are just a few examples of how multiple emerging trends in intelligent communication and wireless technologies are being used collectively to revolutionize various industries. The use of these technologies collectively provides a more intelligent, connected, and automated system, allowing for new applications and services that were not possible before.

7.5 CONCLUSION

In conclusion, this chapter explores the various trends in communication technology and wireless communication with various aspects discussing the applications, benefits, scope of improvements along with industry-specific use cases where multiple trends can be used effectively in an intertwined way. It is clear that the future of intelligent communication and wireless technologies is highly dependent on the growth of several key trends, namely, 5G technology, IoT, AI and ML, and edge computing. It is indicative that the majority of the research community and their research work are driven towards AI as explained in Figure 7.3. It is difficult to determine which of these trends is growing the fastest, as this can vary based on various factors

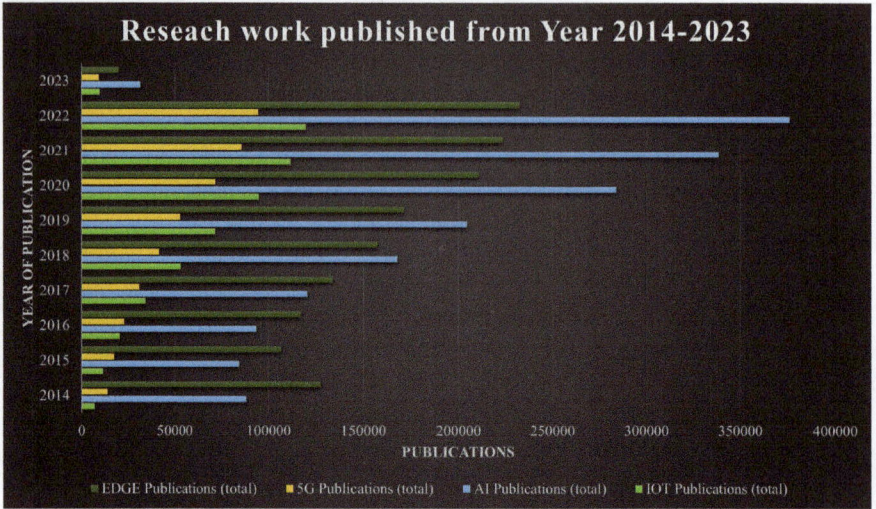

FIGURE 7.3 Research work on different trends published from year 2014 to 2023 (source: https://app.dimensions.ai). This indicates that the research community is more driven towards AI (ML/DL).

such as market demand, technological advancements, and government regulations. However, AI and ML, as well as edge computing, are expected to grow at a rapid pace, driven by the increasing demand for more intelligent and automated systems and more distributed and low latency computing systems, respectively. On the other hand, the 5G market and IoT market are also expected to grow significantly, driven by the increasing demand for higher bandwidth and lower latency in mobile networks and the growing number of 5G applications, services, and connected devices and IoT applications in various industries such as healthcare, manufacturing, and smart cities.

REFERENCES

1. W. de Oliveira *et al.*, "OpenCare5G: O-RAN in private network for digital health applications," *Sensors (Basel).*, vol. 23, no. 2, Jan. 2023, doi: 10.3390/S23021047.
2. Y. Wu, B. H. Yim, C. Lu, L. Mao, and J. J. Zhang, "Can signal delay and advertising lead to profit? A study on sporting," *Front. Psychol.*, vol. 13, Jan. 2023, doi: 10.3389/FPSYG.2022.1028117.
3. J. Ming and R. Qin, "Trends in research related to ophthalmic microperimetry from 1992 to 2022: A bibliometric analysis and knowledge graph study," *Front. Med.*, vol. 9, Jan. 2023, doi: 10.3389/FMED.2022.1024336.
4. D. Wang, A. Qiu, Q. Zhou, S. Partani, and H. D. Schotten, "The Effect of Variable Factors on the Handover Performance for Ultra Dense Network," Jan. 2023, doi: 10.48550/arxiv.2301.08053.
5. X. Zhang, L. Yang, X. Zhang, and J. Xu, "Research on the development trend, evolution, and spatial local characteristics of the intelligent smart medical industry in the Yangtze River Economic Belt," *Front. public Heal.*, vol. 10, Jan. 2023, doi: 10.3389/FPUBH.2022.1022547.

6. Z. Du and T. Wang, "Knowledge domain and dynamic patterns in multimodal molecular imaging from 2012 to 2021: A visual bibliometric analysis," *Medicine (Baltimore).*, vol. 102, no. 4, p. E32780, Jan. 2023, doi: 10.1097/MD.0000000000032780.
7. S. D. Sathyanarayana, M. Sankaradas, and S. Chakradhar, "5GLoR: 5G LAN Orchestration for enterprise IoT applications," Feb. 2023, doi: 10.48550/arxiv.2302.02034.
8. J. C. Lin, "Incongruities in recently revised radiofrequency exposure guidelines and standards," *Environ. Res.*, vol. 222, p. 115369, Apr. 2023, doi: 10.1016/J.ENVRES.2023.115369.
9. S. Fan *et al.*, "Feasibility and safety of dual-console telesurgery with the KangDuo Surgical Robot-01 system using fifth-generation and wired networks: An animal experiment and clinical study," *Eur. Urol. open Sci.*, vol. 49, pp. 6–9, Mar. 2023, doi: 10.1016/J.EUROS.2022.12.010.
10. A. N. Doddamane and A. S. Kumar, "The implications of 5G technology on cardiothoracic surgical services in India," *Indian J. Thorac. Cardiovasc. Surg.*, vol. 39, no. 2, Jan. 2023, doi: 10.1007/S12055-022-01448-6.
11. R. Singh, D. Kukreja, and D. K. Sharma, "Blockchain-enabled access control to prevent cyber attacks in IoT: Systematic literature review," *Front. big data*, vol. 5, Jan. 2023, doi: 10.3389/FDATA.2022.1081770.
12. A. Belhadi, J.-O. Holland, A. Yazidi, G. Srivastava, J. C.-W. Lin, and Y. Djenouri, "BIoMT-ISeg: Blockchain internet of medical things for intelligent segmentation," *Front. Physiol.*, vol. 13, Jan. 2023, doi: 10.3389/FPHYS.2022.1097204.
13. M. U. Alam and R. Rahmani, "FedSepsis: A federated multi-modal deep learning-based internet of medical things application for early detection of sepsis from electronic health records using Raspberry Pi and Jetson Nano Devices," *Sensors (Basel).*, vol. 23, no. 2, Jan. 2023, doi: 10.3390/S23020970.
14. B. Geng and P. K. Varshney, "Human-Machine Collaboration for Smart Decision Making: Current Trends and Future Opportunities," Jan. 2023, doi: 10.48550/arxiv.2301.07766.
15. A. Tselebis, L. Zabuliene, C. Milionis, and I. Ilias, "Pandemic and precocious puberty - a Google trends study," *World J. Methodol.*, vol. 13, no. 1, pp. 1–9, Jan. 2023, doi: 10.5662/WJM.V13.I1.1.
16. E. Yıldırım, M. Cicioğlu, and A. Çalhan, "Fog-cloud architecture-driven Internet of Medical Things framework for healthcare monitoring," *Med. Biol. Eng. Comput.*, 2023, doi: 10.1007/S11517-023-02776-4.
17. A. A. Khan, A. A. Laghari, P. Li, M. A. Dootio, and S. Karim, "The collaborative role of blockchain, artificial intelligence, and industrial internet of things in digitalization of small and medium-size enterprises," *Sci. Rep.*, vol. 13, no. 1, Dec. 2023, doi: 10.1038/S41598-023-28707-9.
18. A. Ghubaish, T. Salman, M. Zolanvari, D. Unal, A. Al-Ali, and R. Jain, "Recent advances in the internet of medical things (IoMT) systems security," *IEEE Internet Things J.*, vol. 8, no. 11, pp. 8707–8718, Feb. 2023, doi: 10.1109/JIOT.2020.3045653.
19. J. M. Kang and D. H. Coelho, "Public health awareness campaigns in otolaryngology: Are we making an impact?," *Ear. Nose. Throat J.*, 2023, doi: 10.1177/01455613221149637.
20. J. Wu, S. Drew, F. Dong, Z. Zhu, and J. Zhou, "Topology-aware federated learning in edge computing: A comprehensive survey," *J. ACM*, vol. 37, no. 1, p. 35, Feb. 2023, doi: 10.48550/arxiv.2302.02573.
21. R. Tornero-Costa, A. Martinez-Millana, N. Azzopardi-Muscat, L. Lazeri, V. Traver, and D. Novillo-Ortiz, "Methodological and quality flaws in the use of artificial intelligence in mental health research: Systematic review," *JMIR Ment. Heal.*, vol. 10, p. e42045, Feb. 2023, doi: 10.2196/42045.

22. T. N. Poly, M. M. Islam, B. A. Walther, M. C. Lin, and Y.-C. (Jack) Li, "Artificial intelligence in diabetic retinopathy: Bibliometric analysis," *Comput. Methods Programs Biomed.*, vol. 231, p. 107358, Apr. 2023, doi: 10.1016/J.CMPB.2023.107358.
23. R. Miller, S. Farnebo, and M. D. Horwitz, "Insights and trends review: artificial intelligence in hand surgery," *J. Hand Surg. Eur.* Vol. 2023, p. 175319342311525, Feb. 2023, doi: 10.1177/17531934231152592.
24. M. Tang *et al.*, "Research frontiers and trends in the application of artificial intelligence to sepsis: A bibliometric analysis," *Front. Med.*, vol. 9, Jan. 2023, doi: 10.3389/FMED.2022.1043589.
25. R. Awasthi *et al.*, "Quantitative and Qualitative evaluation of the recent Artificial Intelligence in Healthcare publications using Deep-Learning," *medRxiv*, p. 2022.12.31.22284092, Jan. 2023, doi: 10.1101/2022.12.31.22284092.
26. G. Joshi, A. Jain, S. Adhikari, H. Garg, and M. Bhandari, "FDA approved Artificial Intelligence and Machine Learning (AI/ML)-Enabled Medical Devices: An updated 2022 landscape," *medRxiv*, p. 2022.12.07.22283216, Jan. 2023, doi: 10.1101/2022.12.07.22283216.
27. S. Huh, "Recent Issues in Medical Journal Publishing and Editing Policies: Adoption of Artificial Intelligence, Preprints, Open Peer Review, Model Text Recycling Policies, Best Practice in Scholarly Publishing 4th Version, and Country Names in Titles," *Neurointervention*, Feb. 2023, doi: 10.5469/NEUROINT.2022.00493.
28. A. Holzinger, K. Keiblinger, P. Holub, K. Zatloukal, and H. Müller, "AI for life: Trends in artificial intelligence for biotechnology," *N. Biotechnol.*, vol. 74, pp. 16–24, May 2023, doi: 10.1016/J.NBT.2023.02.001.
29. K. Oztoprak, Y. K. Tuncel, and I. Butun, "Technological transformation of telco operators towards seamless IoT edge-cloud continuum," *Sensors (Basel).*, vol. 23, no. 2, Jan. 2023, doi: 10.3390/S23021004.
30. H. Liu, L. Tong, M. Su, D. Chen, G. Song, and Y. Zhou, "The latest research trends in the removal of cesium from radioactive wastewater: A review based on data-driven and visual analysis," *Sci. Total Environ.*, vol. 869, Apr. 2023, doi: 10.1016/J.SCITOTENV.2023.161664.
31. L. F. S. de Castro and S. Rigo, "Relating Edge Computing and Microservices by means of Architecture Approaches and Features, Orchestration, Choreography, and Offloading: A Systematic Literature Review," Jan. 2023, doi: 10.48550/arxiv.2301.07803.
32. G. Ming, "Exploration of the intelligent control system of autonomous vehicles based on edge computing," *PLoS One*, vol. 18, no. 2, Feb. 2023, doi: 10.1371/JOURNAL.PONE.0281294.
33. F. Tian, J. Yang, S. Zhao, and M. Sawan, "NeuroCARE: A generic neuromorphic edge computing framework for healthcare applications," *Front. Neurosci.*, vol. 17, Jan. 2023, doi: 10.3389/FNINS.2023.1093865.
34. E. J. Cornblath *et al.*, "Quantifying trial-by-trial variability during cortico-cortical evoked potential mapping of epileptogenic tissue," *Epilepsia*, Feb. 2023, doi: 10.1111/EPI.17528.

8 Internet of Multimedia Things (IoMT)

Communication Techniques Perspective

Vinay Kumar, Sumit Kushwaha,
Indrasen Singh, Rabindra K Barik,
Gyanesh Singh, and Manish Sabraj

8.1 INTRODUCTION

Internet of Multimedia Things (IoMT) is envisioned as one of the integrated models with multimedia and the Internet of Things and has been considered ab advanced technology in the present scenario [1, 2]. The IoMT is associated with the devices/things all around us, like mobile devices, radio frequency identification (RFID) devices, different actuators, and sensors with the internet through distant association or wired parameters. Consequently, it allows things to connect with their neighbors to improve the desired systems or structures [2, 3]. A couple of experts have portrayed IoMT in various settings [2] as the joining of tiny gadgets known as smart objects (SO); for the most crucial part, battery works outfitted with the various microcontroller and handsets into network devices for innovative services [3–6]. IoMT has set out new open entryways for machines to talk with each other and expand recommended applications. Presently, 23 billion gadgets are linked with cyberspace, and these numbers will increase to 40 billion within the next five years [7, 8]. Typically, IoMT has retained for machine-to-machine (M2M), human-to-machine (H2M), machine in/or humans (MiH), and human-to-human (H2H) ecosystem in statistics with distinguishing gadgets [9]. Quick improvement in the related gadgets to cyberspace during the latest decade and unexpected interest for blended-media traffic have offered a climb to the ascent of the IoMT [10]. The concordance between quality of service (QoS) data is represented towards an augmentation in intuitive media for QoS data.

IoMT generates an enormous volume of information with different characteristics and necessities than the Internet of Things (IoT). The real-time, application-based scenarios from intelligent traffic monitoring to intelligent hospitals has been discussed [11]. Hence, timely decisions and delivery of information in IoMT are critical as they directly relate to human beings' safety. IoMT devices are different from IoT devices. They require massive storage and excessive processing, and are more power hungry with higher bandwidth.

This chapter aims to summarize the current issues of IoMT from various communication techniques' perspectives. We review the communication technologies for IoMT, key features of IoMT data, and provide comparative analysis along with the future scope.

8.2 COMMUNICATION TECHNOLOGIES FOR IOMT

The fast growth of the world's population demands more food, natural resources, and space to reside. This ever-increasing demand requires novel technologies to explore resources and outer space. Enabling technology to overcome the above concerns mostly comes under the Internet of X-things (X-IoT) [12–16]. The framework of X-IoT consists of three major categories: terrestrial (conventional/space), underwater, and underground (non-conventional). The first category is the Internet of Terrestrial Things (IoT-TS) for outer space exploration, to provide coverage, and to enable inter-satellite communications. The second category is the Internet of Underwater Things (IoT-UW) for ocean exploration. The third category is the Internet of Underground Things (IoT-UG) for agriculture, seismic monitoring, and oil and gas exploration [17–20].

The technological challenges for IoT-TS, IoT-UW, and IoT-UG can be broadly classified into communication, networking, and localization. For conventional IoTs (IoT-ST), radiofrequency communication (RF) is the best technique for transmitting information between devices [21]. For non-conventional media like underwater and underground IoT applications, traditional communication techniques like electromagnetics (RF) are not appropriate, mainly because of high path loss and thus being unfeasible. Most of the research is based on acoustic communication in underwater media, which means the sound wave transfers information. However, it faces many challenges like bad channel behavior (dynamic), low data rate, and significant delay. Acoustic-based communication provides a long transmission range at a lower frequency, but the data rate is very low due to the slower speed of sound waves transversal. Therefore, optical waves-based communication has been introduced to provide a large data rate for underwater environments [22–26]. However, this type of communication underwater also suffers from low communication range and requires line-of-sight (LOS) communication. Magnetic induction (MI)-based communication is a recently explored technique for underwater and underground applications. This promising communication technique exhibits several characteristics such as constant channel behavior, very low propagation delay of high speed, large communication range (up to 100 meters) with a sufficiently large data rate [27–31].

The shortage of internet protocol version 4 (IPv) desires the suggestion of next-generation IPv6 for use of oMT gadgets. IPv6 with low power personal area networks (LoWPAN) is an IP-built innovation for Lossy-Networks and low power to overcome any issues between next-generation IP world and low-power gadgets [32]. IPv6 with LoWPAN packs next-generation-IPv6 headers and discontinuity of huge parcels to make next-generation IPv6 appropriate for source-controlled

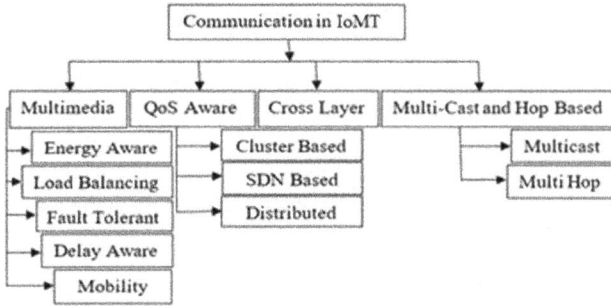

FIGURE 8.1   Classifications of communications in IoMT.

gadgets [33]. The International Engineering Task Force (IETF) proposed a next-generation IPv6 direction-finding protocol for Lossy-Networks [34, 35]. This is currently broadly received as a hopeful norm with expanding ubiquity since it proposes adaptability and interoperability by embracing diverse organization geographies. routing protocol for low-power and lossy networks (RPL) abuses key organization measurements like hop-count, throughput, node energy, latency, and link consistency. This next-generation IPv6 direction-finding protocol for Lossy-Networks upholds person-to-person (P2P), person-to-group-of-persons (P2MP), and group-of-person-to-person (MP2P) communication worldviews. Authors [36] gave a definite image of the exploration that has examined RPL. The creators introduced the circulation of examination completed on RPL regarding the topographical district, working frameworks, equipment stages, and organization measurements [25, 37–40]. We have broadly considered IoMT communication to meet tough prerequisites of interactive media communication in IoMT while keeping up Quality of Experience (QoE) and QoS with adequate data transfer capacity and energy use. Figure 8.1 shows the classifications of communication in IoMT.

We can deploy the IoMT devices on terrestrial places, underwater, and underground. Details of these IoMT devices are classified into terrestrial IoMT, underground IoMT, and underwater IoMT, which are described in detail in next the section [41]. **IOT-TS:** Terrestrial IoMT devices are a group of a few remote sensor nodes [42]. These nodes are capable of communicating with base stations effectively and include of hundreds to thousands of remote sensor nodes placed either in organized or unstructured ways. In unstructured mode, the sensor node is randomly distributed inside the objective territory that is slumped from a fixed plane. The organized or preplanned mode thinks about optimal 2D with 3D models deployment and grid deployment models. The major drawback with these nodes is battery power, which is limited. However, the battery is equipped with solar cells as a secondary source for power. The energy conservation of these sensor nodes is accomplished by utilizing low duty cycle tasks, minimizing-delays, and optimal-routing, etc. Figure 8.2 indicates the examination of key information includes among IoT and IoMT

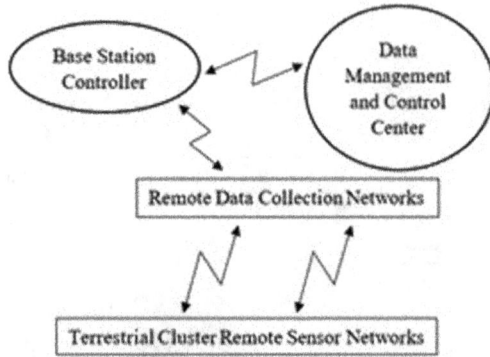

FIGURE 8.2   Terrestrial remote sensor networks.

8.2.1   UNDERGROUND IoMT DEVICES

Underground IoMT devices are a group of a unique kind of remote sensor node. Underground remote sensor nodes are more costly than earthly remote sensor nodes regarding placement, maintenance, and equipment cost, and require careful planning. Underground remote sensor nodes are wrapped around to examine underground conditions. To relay information from the sensor nodes to the base station, more sink nodes are situated over the ground. The underground remote sensor nodes that are sent into the ground are difficult to re-energize. The sensor node batteries equipped with a limited-battery power are hard to re-energize. Likewise, the underground environment becomes remote communication as a challenge due to the high level of attenuation and signal loss. Underground remote sensor networks is shown in Figure 8.3.

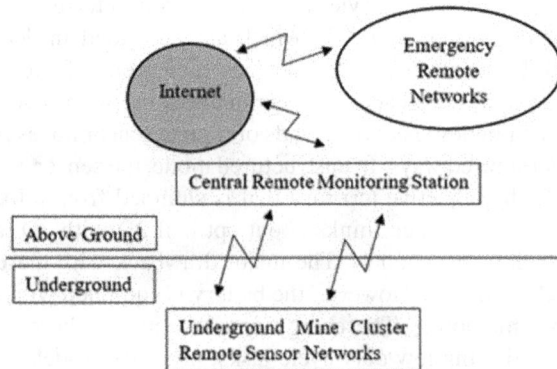

FIGURE 8.3   Underground remote sensor networks.

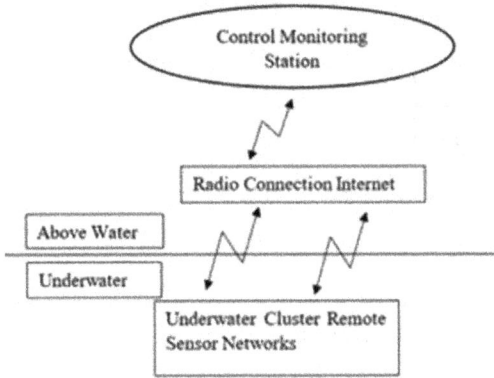

FIGURE 8.4 Underwater remote sensor networks.

8.2.2 UNDERWATER IoMT DEVICES

Over 70 percent of the earth is covered with water. Underwater IoMT devices are a group of a particular type of remote sensor nodes. These IoMT devices are comprised of a few sensor nodes and vehicles placed underwater. Independent underwater vehicles are deployed for collecting information from these sensor nodes. Long propagation delays, high bandwidth requirements, and sensor failures are many challenges in underwater communication. Underwater, these remote sensor nodes are placed with a limited battery capacity that can't be re-energized or replaced. The major issues for underwater remote sensor nodes include the development of underwater communication, networking systems, energy preservation, etc. Figure 8.4 shows the underwater remote sensor networks.

8.3 FEATURES OF IOMT

IoMT devices are unique in relation to IoT devices. They entail greater memory, higher computational force, and with higher transfer speed [43]. The ongoing setup scenarios range from smart residences, clever horticulture, smart emergency clinics, smart urban neighborhoods, and astute frameworks. The principle for IoMT is the convenient and dependable conveyance of the information. Hence, it forces severe QoS necessities and requests productive organization engineering. The client's point of view of QoS is identified as QoE. QoE can be additionally described as evenhanded or abstract. The client's goal for QoE is to quantify it, and it changes dramatically depending on what's needed. Nonetheless, specialist organizations worry with the abstract QoE to assess the organization mean assessment score. The sight and sound information are expanding multifold. It presents new difficulties to communicate, cycle, store and offer the information. Handling involves brand new procedures for edge, haze, and cloud assisted devices. Additional pressure and decompressing procedures are presented for the capacity of interactive media information. Directing convention for low-power and lossy organizations

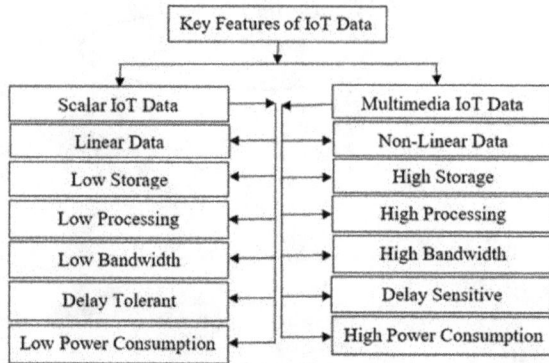

FIGURE 8.5   Comparison of key data features between IoT and IoMT.

is the standard IoT steering convention. It demands further improvement by contemplating energy-mindful, load adjusting, adaptation to non-critical failure, and defer mindful IoMT sending situations [44, 45]. Comparison of key data features between IoT and IoMT is provided in Figure 8.5. Further, comparative evaluation of RMNS, IoT and IoMT over different parameters is given in Table 8.1.

8.4   CASE STUDY OF TERRESTRIAL IOMT

In this section, we have taken one example of terrestrial IoMT. Video communication has become essential in our daily life. The usage of video sensor networks is also increasing day by day. For a given environment, either with wired or wireless

TABLE 8.1

Evaluation of RMNS, IoMT, and IoT with Respect to Various Characteristics

Parameters	Remote Multimedia Sensor Networks (RMSN)	IoT	IoMT
Application Scope	Deployment Dependent	Flexible and Dynamic	Flexible and Dynamic
Process of Nodes	Pre-defined	Adaptive	Adaptive
Capabilities and Operation of Nodes	Homogeneous	Heterogeneous	Heterogeneous
Energy Efficient	Yes	Yes	Yes
Scalability	Extreme	Extreme	Extreme
Interoperability	Moderate	Extreme	Extreme
Cost of Deployment	Low Cost	Low Cost	Low Cost
Topology	Limited	Dynamic and Ad Hoc	Dynamic and Ad Hoc
Capacity of Bandwidth	Modest	Low	Low
Connectivity of IPs	Limited	Uniquely Addressable	Uniquely Addressable
QoS Multimedia	Limited	Unreachable	Reachable

modes of communication, the quality of service depends on factors like bandwidth, computing power, energy backup, and memory and so on. A video communication requires high bandwidth to achieve better quality of service in terms of video resolution. Further, for a secure and real-time communication, bandwidth and computing power of a node play a critical role in the networks. For uninhabited environments (such as hill areas, forest areas, line of control on borders) the challenge is multi-fold. In these environments mostly the areas are remote in nature and backbone infrastructure of network is not available. Establishing backbone network in these areas is difficult and costly. Therefore, ad hoc networks are preferred for communicating in such areas. Most of the protocols deployed in ad hoc networks such as Wi-max, ZigBee, and Bluetooth, etc., have their own constraints. These protocols provide limited bandwidth along with constrained computing capacity. The video sensor nodes invariably have limitation in terms of battery power. Providing effective and secure video communication for such areas becomes very challenging because the video data to be sent in real time needs high bandwidth and moreover, providing security will result in more redundant data. This also requires more computing power which leads to more energy consumption. In view of this, we have proposed a network model as given in below Figures 8.6 and 8.7.

In the proposed model, the video sensor nodes equipped with ZigBee modules are deployed. WiMAX nodes are also deployed (static/dynamic). The network is divided into clusters where WiMAX nodes serve as cluster head (CH) and ZigBee nodes serve as cluster members (CM). The video sensor nodes will communicate with WiMAX nodes. The WiMAX node will receive data from the CMs. Data will be aggregated at WiMAX node and will be forwarded to base station which is

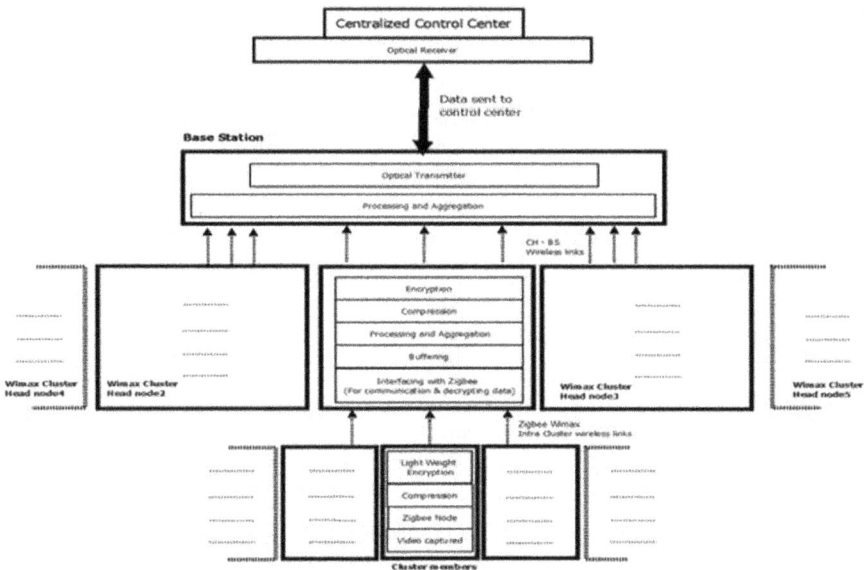

FIGURE 8.6   Architecture of ZigBee-WiMAX interfaced video sensor network.

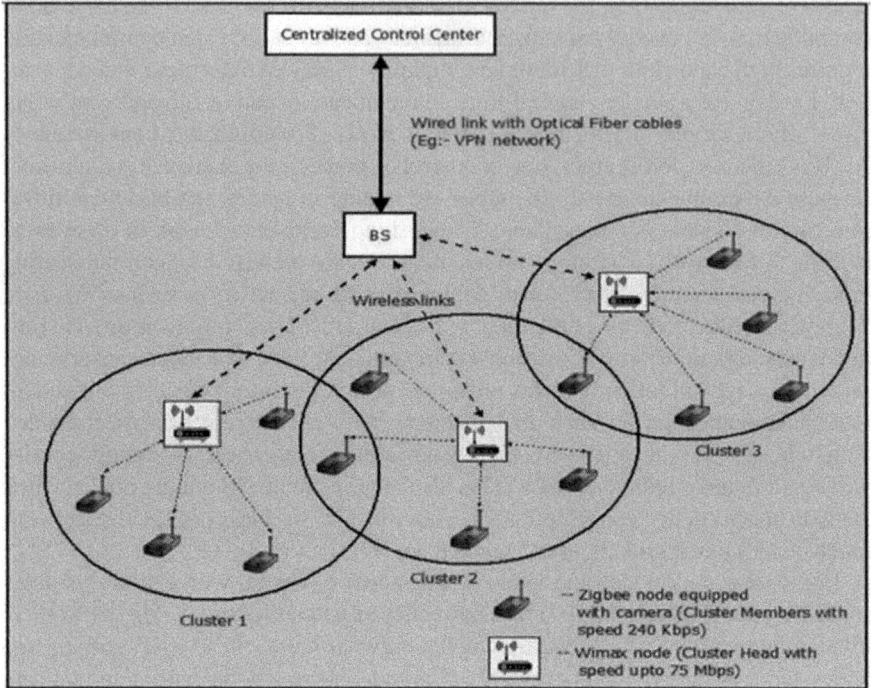

FIGURE 8.7   Clustered sensor network for secure real-time video processing.

securely connected to backbone network. On the other hand, the WiMAX protocol provides bigger range for communication along with higher bandwidth. The data received from the ZigBee node is buffered at WiMAX Node and processing will be done on data to remove redundant data. Further, the data aggregation technique will be applied before applying compression and encryption techniques. For getting higher data rate without changing the input power level, a multiple-input multiple-output MIMO technology is widely used, in which multiple antennas are placed at transmitter and receiver side [46–48]. Spatial diversity and spatial multiplexing techniques provide higher data rates. Spatial diversity enhances the data rate by improving the link reliability. On the other hand, spatial multiplexing enhances the data rate by sending the encoded data stream separately with different users [49]. Different researchers are working in different areas of MIMO antenna design, such as bandwidth enhancement and betterment of isolation. Most researchers are focused on reduction of mutual coupling between closely spaced radiators. Different mutual coupling reduction techniques available in literature such as use of parasitic elements [50], use of defected ground structure (DGS) [51].

Polarization diversity among the ports and use of neutralization line has been discussed [52]. Currently, to avoid the problem of low gain value, some research work has also been done in the field of dielectric resonator (DR)-based MIMO antennas. Sharma et. al. proposed a tri-band hybrid dielectric resonator-based MIMO antenna.

It consists of two ports, and isolation between the ports has been enhanced by using stub at the ground plane [53].

Data at ZigBee node can be pre-processed to adjust the data rate by reducing the redundant information. This will involve differential coding to eliminate the static information between the video frames and adjusting the frame rate for required data rate and acceptable quality of service [53]. Clustering is a technique that reduces energy consumption and provides proper traffic load distribution in sensor networks. In clustering technique, sensor nodes form clusters and each cluster has one cluster head (CH) node and many non-CH nodes. Non-CH nodes send the data to the CH and CH transmits the data to BS through other CH. But many times it happens that non-CH nodes in a cluster send similar information to the CH. Hence, it creates a large amount of redundant information at the CH node. This redundant information must be removed before transmitting to base station (BS) to increase the network lifetime. To implement data buffering along with data aggregation and processing at WiMAX node, various data aggregation techniques are available but the challenges to select and implement are the same in the proposed model [54]. The effort will be laid down to investigate these techniques.

The ZigBee protocol used within the cluster members has low data rate and low transmission range. The video data captured at video sensors (using a 2-megapixel camera) will be compressed and encrypted (using lightweight encryption technique) before sending to WiMAX node. Further, it is to be noted that the data compression technique will be applied at video sensor nodes to reduce the size of video data. Normally a 2-megapixel camera captures 19–20,000 pixels per frame and every pixel is represented by 4 bits, thus resulting in 7500 Kb data. After performing compression this data size per frame reduces to 37 Kb data. Now this frame can be transferred with ZigBee protocols to WiMAX node with acceptable QoS. Additionally, we are also planning to implement the MIMO concept, which will increase the data rate and link reliability. To achieve interfacing objective the protocol stack of ZigBee and WiMAX need to be customized. The WiMAX protocol stack will be implemented to receive the data from the ZigBee node and further other operations such as data aggregation compression and security will be integrated. The protocol associated with the node discovery and authorization of ZigBee nodes will be developed. Implementation of security services at WiMAX standard security services such as confidality, integrity, and authorization will be implemented over data at WiMAX node. In order to make the information secure, an energy efficient computing techniques need to be developed.

8.5   APPLICATIONS OF IOMT

Currently, IoMT is working in different areas such as M2M, H2M, H2H, and in MiH trades in numerous areas such as the geospatial sector, clinical checking tests, and healthcare sector [10]. IoMT splendid things are usually resource constricted, similar to memory amassing, planning power, and energy. To produce the gadgets more humble and energy-gainful, sensors are ordinarily proposed to be battery worked or daylight based filled two or three KB of memories and restricted dealing with power in megahertz. Media QoS data show contradictory conduct when it appears differently about representative IoMT scalar data. IoMT-related gadgets involve higher exchange speeds, higher computational capacity, monstrous memory resources, and taking apart and handling the acquired media data. The standard intuitive media application incorporates the data communication of feature point, feature multipoint, or multipoint to multipoint. Despite the customarily desired, IoMT related applications involve tremendous data transmission during point-to-point correspondence (e.g., the observation course of action of the entire smart city) or multipoint-to-multipoint. Heterogeneous data, better throughput, vibrant structure, QoS, and concede affectability over such resource obliged IoMT adroit things to increase IoMT's troubles. Intuitive media statistical data (i.e., sound, picture, and video) are a set of amorphous features. The extension of intuitive media statistical data acquiring and correspondence involves alteration and improvement of the standard IoT structure, which implies the concept of IoMT. The change of the IoMT concept for media correspondence necessitates capable part event dealing, encoding/deciphering, extraction, energy careful count, lightweight and need-functioning coordination, QoE, and QoS, which is keeping up execution estimations, convincing channel access, and sensible media access control (MAC) shows. Genuine interactive media applications incorporate an illustration of salvage vehicles-based crisis reaction frameworks, traffic observing, GPS-based way following, farming checking, wrongdoing assessment, brilliant urban communities, smart homes, shrewd historical center, reconnaissance frameworks, security framework for confirmation and approval, sight and sound based e-wellbeing, patient checking in keen medical clinics, and modern checking frameworks. Figure 8.8 shows the abstract of IoMT in each area's precise application.

IoT features support multimedia communications. In such cases, video, image, and audio systems are delay-sensitive and data bandwidth hungry. The fast development of image, audio, and video traffic in IoT needs to drive the best approach to enhance new procedures to meet its prerequisites. IoMT devices involve higher transmission capacity, more significant memory, and quicker computational assets to handle information. Ordinary correspondences incorporate multipoint-to-point and multipoint-to-multipoint situations. Genuine interactive media applications incorporate crisis reaction frameworks, traffic observing, wrongdoing investigation, developed urban

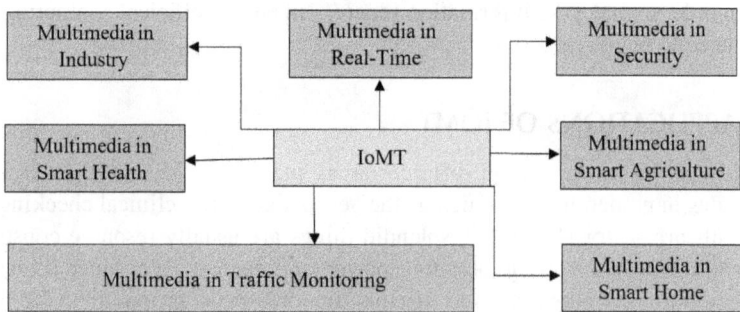

FIGURE 8.8   Abstract of IoMT in each area.

areas, advanced homes, smart clinics, advanced farming, observation frameworks, Industrial IoT (IIoT), and Internet of bodies (IoB). Dynamic organizations, diverse devices, and information, exacting QoS, postponed affectability and unwavering quality necessities around assets compelled IoMT to present enormous difficulties for media correspondence in IoT. Organization on-chip design [44, 45] is one of the practical answers for improving the client nature of involvement.

8.6 CONCLUSION

The emerging IoMT has encouraged several pioneering applications and planning to enhance user satisfaction by associating various smart gadgets/devices with empowering innovations. Multimedia communication in an IoT environment can sustain countless applications. The present book chapter aims to feature the outline of IoMT, the significance of IoMT applications, and an overview of its communication perspectives. While planning IoMT gadgets, architectures, protocols, and computing methods are investigated to give efficient IoMT engineering to help the QoE. Different application usage case models are introduced to explain how IoMT devices/gadgets can reform cyberspace. Finally, this book chapter brings up the wide open and potential research challenges that should be addressed in upcoming IoMT environments. This book chapter will also provide a basis for all researchers trying to explore and understand the challenges and issues with IoMT devices/gadgets. Figure 8.9 shows future scope in IoMT system.

Future Scope
- IoMT gadgets produce bulk data which include labeled and unlabeled data. The processing and handling of such an amount of unstructured information over an energy deficient network necessitates the transmission of only

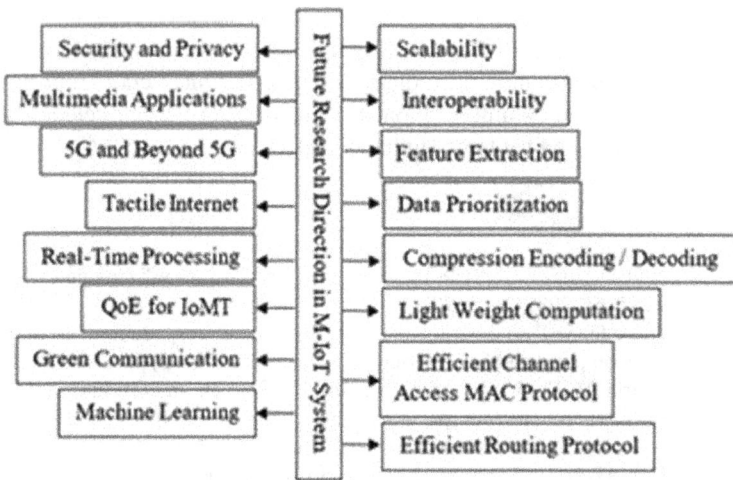

FIGURE 8.9   Future scope in IoMT system.

valuable data. An intelligent feature extraction technique must be required to detect and classify multimedia information with minimal energy consumption.

- Arranging IoMT data represents several issues subjected to video streaming, power, scheduling, improving PLR, memory, enhancing service quality, and reducing network delay and security. Multimedia applications to transfer video, audio, and another form of information necessitate pervasive transmission. Artificial intelligence-based data classification and priority optimization in IoMT should be broadly studied concerning the necessities of the applications.

- The extensive data contents are being recorded in multimedia gadgets, of which video on YouTube, shows off TV, and animations cover 60–70 percent of the internet traffic. A smart/intelligent event mining images, audio, and videos are required in IoMT for effective resource utilization to process the required service with the lowest delay [55].

- No smart/intelligent effort has been done to transmit multimedia data like images, audio, and video over the IoMT network. There is a necessity to search the coding procedures related to multimedia services in IoMT while bearing in mind the power and interface constraints in the system.

- Multimedia processing depends on efficient routing, compression, MAC protocols, network coding, etc. However, little work has been done in software-defined radio as multimedia computation in IoT. To increase network throughput (actual output) with minimum energy utilization, the intelligent transmission of only meaningful information to cloud and fog/edge is required for processing.

- IoMT gadgets support various applications, but IoMT gadgets' security is still a significant issue [56].

- High data transmission and high bandwidth availability are still yet to be explored and examined for integrating IoMT gadgets with 5G and 6G technologies [57–59].

- Available possibilities of research domain from the perspective of nano things, molecular communication, remote physical interaction, robotic surgery, human bond communication, automation industry, and teleoperation are still present and yet to be studied in detail [22–26, 42, 60–62].

REFERENCES

1. A. Luigi and M. Giacomo, "The Internet of Things: A Survey," Computer Networks, vol. 54, no. 15, pp. 2787–2805, 2010.
2. P. P. Ray, "A Survey on Internet of Things Architectures," Journal of King Saud University - Computer and Information Sciences, vol. 30, no. 3, pp. 291–319, 2018.
3. Y. Bin Zakaria, H. Yu, M. K. Afzal, M. Husain Rehmani and O. Hahm, "Internet of Things (IoT): Operating Systems, applications and Protocols Design, and Validation Techniques," Future Generation Computer Systems, vol. 88, pp. 699–706, 2018.
4. S. A.Alvi, B. Afzal, G. A. Shah, L. Atzori and W. Mahmood, "Internet of multimedia things: Vision and Challenges," Ad Hoc Networks, vol. 33, pp. 87–111, 2015.

5. M. Novak, "Pale o future," [Online]. Available: https://paleofuture.gizmodo.com/ nikola-teslas-incredible-predictions-for-ourconnected-1661107313. [Accessed on 9/3/2023].

6. T. Harwood, "Postscapes," [Online]. Available: https://www.postscapes.com/iot-history/. [Accessed on 9/3/2023].

7. "Statista," [Online]. Available: https://www.statista.com/statistics/471264/iot-numberof-connected-devices-worldwide/. [Accessed on 9/3/2023].

8. "IEEE Spectrum," [Online]. Available: https://spectrum.ieee.org/popular-internet-of-things-forecast-of-50-billion-devices-by-2020-is-outdated. [Accessed on 9/3/2023].

9. S. Kushwaha, "Review on Artificial Intelligence and Human Computer Interaction," IEEE OPJU International Technology Conference on Emerging Technologies For Sustainable Development, organized by O.P. Jindal University, Raigarh, Chhattisgarh, India, pp. 1–6, 2023.

10. D. Minoli, Building the Internet of Things with IPv6 and MIPv6, New Jersey: Wiley & Sons, 2013.

11. ITU, "Traffic Routing Recommendation E.170," International Telecommunication Union, 2007.

12. K. Sood, S. Yu and Y. Xiang, "Software-Defined Wireless Networking Opportunities and Challenges for Internet-of-Things: A Review," IEEE Internet of Things Journal, vol. 3, no. 4, pp. 453–463, 2016.

13. D. Mera, M. Batko and P. Zezula, "Towards Fast Multimedia Feature Extraction: Hadoop or Storm," in 2014 IEEE International Symposium on Multimedia, Taichung, 2014.

14. Q. Wang, W. Wang and K. Sohraby, "Multimedia Relay Resource Allocation for Energy Efficient Wireless Networks: High-Layer Content Prioritization With LowLayer Diversity Cooperation," IEEE Transactions on Vehicular Technology, vol. 66, no. 11, pp. 10394–10405, 2017.

15. L. Fan, H. Du, U. Mudugamuwa and B. G. Evans, "A Cross-Layer Delay Differentiation Packet Scheduling Scheme for Multimedia Content Delivery in 3G Satellite Multimedia Systems," IEEE Transactions on Broadcasting, vol. 54, no. 4, pp. 806–815, 2008.

16. M. Jafari and S. Kasaei, "Prioritisation of Data Partitioned MPEG-4 for Streaming Video in GPRS Mobile Networks," 2005 1st IEEE and IFIP International Conference in Central Asia on Internet, Bishkek, 2005.

17. V. Kumar and S. Kushwaha, "Map-Reduce Task Scheduling Optimization Techniques: A Comparative Study," 2023 7th International Conference on Trends in Electronics and Informatics (ICOEI), Tirunelveli, India, pp. 729–736, 2023.

18. A. Iera, A. Molinaro, G. Ruggeri and D. Tripodi, "Dynamic Prioritization of Multimedia Flows for Improving QoS and throughput in IEEE 802.11e WLANs," IEEE International Conference on Communications, 2005. ICC, Seoul, 2005.

19. R. Gnanajeyaraman, M. Ganeshkumar and C. Manojkumar, "Performance and Analysis of Prioritization Security for Input-Queued Packet Switches," 2010 International Conference on Communication and Computational Intelligence (INCOCCI), Erode, 2010.

20. M. van der Schaar and D. S. Turaga, "Cross-Layer Packetization and Retransmission Strategies for Delay-Sensitive Wireless Multimedia Transmission," IEEE Transactions on Multimedia, vol. 9, no. 1, pp. 185–197, 2007.

21. S. Raviraja, K. Seethalakshmi, S. Kushwaha, P. M. Vishnu Priya, K. R. Kumar, and B. Dhyani, "Optimization of the ART Tomographic Reconstruction Algorithm - Monte Carlo Simulation," 2023 2nd International Conference on Applied Artificial Intelligence and Computing (ICAAIC), Salem, India, pp. 984–988, 2023.

22. E. Steinbach, et al., "Haptic Codecs for the Tactile Internet," Proceedings of the IEEE, vol. 107, no. 2, pp. 447–470, 2019.

23. Y. Sheng, "Scalable Intelligence-Enabled Networking with Traffic Engineering in 5G Scenarios for Future Audio-Visual-Tactile Internet," IEEE Access, vol. 6, pp. 30378–30391, 2018.

24. S. Sukhmani, M. Sadeghi, M. Erol-Kantarci and A. El Saddik, "Edge Caching and Computing in 5G for Mobile AR/VR and Tactile Internet," IEEE MultiMedia, vol. 26, no. 1, pp. 21–30, 2019.

25. S. Dixit, S. Mohan, R. Prasad and H. Harada, "Multi-Sensory Human Bond Communication," IEEE Communications Magazine, vol. 57, no. 2, pp. 18–18, 2019.

26. I. Singh and L. S. Won "Self-Adaptive and Secure Mechanism for IoT Based Multimedia Services: A Survey," Multimedia Tools and Applications, vol. 81, pp. 1–36, 2021.

27. S. Yadav, V. Kumar, S. B. Dhok and D. N. Jayakody, "Energy-Efficient Design of MI Communication-Based 3-d Non-Conventional WSNS," IEEE Systems Journal, vol. 14, no. 2, pp. 2585–2588, 2019.

28. L. Prasanna, V. Kumar and S. B. Dhok, "Cooperative Communication and Energy-Harvesting-Enabled Energy-Efficient Design of Magnetic Induction-Based Clustered NonconventionalWSNs," IEEE Systems Journal, Published Online, pp. 2293–2302, 2019

29. V. Kumar, R. Bhusari, S. B. Dhok, A. Prakash, R. Tripathi and S. Tiwari, "Design of Magnetic Induction Based Energy-Efficient WSNs for Nonconventional Media Using Multilayer Transmitter-Enabled Novel Energy Model," IEEE System Journal, vol. 99, pp. 1932–8184, 2018.

30. A. K. Sharma, S. Yadav, S. Dandu, V. Kumar, J. Sengupta, S. B. Dhok and S. Kumar, "Magnetic Induction-Based Non- Conventional Media Communications: A Review," IEEE Sensor Journal, vol. 17, no. 4, p. 926

31. S. Tambe, Viny Kumar and R. Bhushari, "Magnetic Induction based Cluster Optimization inNon-conventional WSNs: A Cross Layer Approach," International Journal of Electronics and Communication Engineering (Elsevier), vol. 98, pp. 5362, 2018.

32. G. Montenegro, N. Kushalnagar, J. Hui and D. Culler, "Transmission of IPv6 Packets over IEEE 802.15.4 Networks," Internet Proposed Standard RFC 4944, 2007.

33. N. Kushalnagar, G. Montenegro and C. Schumacher, "IPv6 over Low-Power Wireless Personal Area Networks (6LoWPANs): Overview, Assumptions, Problem Statement, and Goals," IETF RFC 4919, 2007.

34. T. Winter, P. Tnodeert, A. Brandt, J. Hui, R. Kelsey, P. Levis, K. Pister, R. Struik, J. Vasseur and R. Alexander, "RPL: IPv6 Routing Protocol for Low-Power and Lossy Networks," Internet Engineering Task Force (IETF) RFC 6550, 2012.

35. O. Gaddour and A. Koubaa, "RPL in a nutshell: A survey," Computer Networks, vol. 56, no. 14, pp. 3163–3178, 2012.

36. H. S. Kim, J. Ko, D. E. Culler and J. Paek, "Challenging the IPv6 Routing Protocol for Low-Power and Lossy Networks (RPL): A Survey," IEEE Communications Surveys & Tutorials, vol. 19, no. 4, pp. 2502–2525, 2017.

37. I. Ishaq, D. Carels, G. K. Teklemariam, J. Hoebeke, F. V. d. Abeele, E. D. Poorter, I. Moerman and P. Demeester, "IETF Standardization in the Field of the Internet of Things (IoT): A Survey," Journal of Sensor and Actuator Networks, vol. 2, pp. 235–287, 2013.

38. T. Salman and R. Jain, "Networking protocols and standards for internet of things," in Internet of Things and Data Analytics Handbook, John Wiley & Sons, Inc., 2017.

39. Y. B. Zikria, H. Yu, M. K. Afzal, M. H. Rehmani and O. Hahm, "Internet of Things (IoT): Operating System, Applications and Protocols Design, and Validation Techniques," Future Generation Computer Systems, vol. 88, pp. 699–706, 2018.

40. A. Al-Fuqaha, M. Guizani, M. Mohammadi, M. Aledhari and M. Ayyash, "Internet of Things: A Survey on Enabling Technologies, Protocols, and Applications," IEEE Communications Surveys & Tutorials, vol. 17, no. 4, pp. 2347–2376, 2015.

41. [Online]. Available: https://www.elprocus.com/architecture-of-wireless-sensor-network-and-applications/ [Accessed on 9/3/2023].

42. S. Kushwaha, V. Kumar and S. Jain, "Node Architectures and Its Deployment in Wireless Sensor Networks: A Survey," SPRINGER International Conference on HPAGC - 2011, CCIS 169, pp. 522–533, 2011.

43. Y. Meng, M. A. Naeem, R. Ali, Y. B. Zikria and S. W. Kim, "DCS: Distributed Caching Strategy at the Edge of Vehicular Sensor Networks in Information-Centric Networking," Sensors, vol. 2019, p. 4407, 2019.

44. Y. A. Qadri, A. Nauman, Y. B. Zikria, A. V. Vasilakos and S. W. Kim, "The Future of Healthcare Internet of Things: A Survey of Emerging Technologies," IEEE Communication Survey and Tutorial, vol. 22, no. 2, p. 1121–1167, 2020.

45. M. Ibrahim, N. K. Baloch, S. Anjum, Y. B. Zikria, S. W. Kim, An Energy Efficient and Low Overhead Fault Mitigation Technique for Internet of Thing Edge Devices Reliable on-Chip Communication. Software: Practice and Experience. Available online: https://onlinelibrary.wiley.com/doi/pdf/10.1002/spe.2796 (Accessed on 9/3/2023).

46. D. N. Sandeep and V. Kumar, "Review on Clustering, Coverage and Connectivity in Underwater Wireless Sensor Networks: A Communication Techniques Perspective," IEEE Access, vol. 5, pp. 11176–111999, 2017

47. S. Yadav and V. Kumar, "Optimal Clustering in Underwater Wireless Sensor Networks: Acoustic, EM and FSO Communication Compliant Technique," IEEE Access, vol. 5, pp. 2169–353, 2017

48. Y. B. Zikria, M. K. Afzal and S. W. Kim, "Internet of Multimedia Things (IoMT): Opportunities, Challenges and Solutions," Sensors, vol. 20, no. 8,p. 2334, 2020.

49. A. Lozano and N. Jindal, "Transmit Diversity vs. Spatial Multiplexing in Modern MIMO Systems," IEEE Transactions on Wireless Communications, vol. 9, pp. 186–197, 2010.

50. S. Zhang, Z. Ying, J. Xiong, and S. He, "Ultra-wideband MIMO/diversity antennas with a tree-like structure to enhance wideband isolation," IEEE Antennas and Wireless Propagation Letters, vol. 8, pp. 1279–1282, 2009.

51. L. Liu, S. W. Cheung, and T. I. Yuk, "Compact MIMO Antenna for Portable Devices in UWB Applications," IEEE Transactions on Antennas and Propagation, vol. 61, no. 8, pp. 4257–4264, Aug. 2013.

52. T.-C. Tang, and K.-H. Lin, "Ultra-Wideband MIMO Antennas with Dual Band-Notched Function," IEEE Antennas and Wireless Propagation Letters, vol. 13, pp. 1076–1079, 2014.

53. A. Sharma, G. Das, R. K. Gangwar, "Design and Analysis of Tri-Band Dual-Port Dielectric Resonator Based Hybrid Antenna for WLAN/WiMAX Applications," IET Microwaves Antennas & Propagation, vol. 12(6), pp. 986–992, 2018.

54. G. Das, A. Sharma, R. Kumar Gangwar and M. S. Sharawi, "Performance Improvement of Multiband MIMO Dielectric Resonator Antenna System with a Partially Reflecting Surface," IEEE Antennas and Wireless Propagation Letters, vol. 18, no. 10, pp. 2105–2109, 2019.

55. S. Kushwaha, "A Futuristic Perspective on Artificial Intelligence," IEEE OPJU International Technology Conference on Emerging Technologies For Sustainable Development, organized by O.P. Jindal University, Raigarh, Chhattisgarh, India, pp. 1–6, 2023.

56. A. M. Mezher and M. A. Igartua, "Multimedia Multimetric Map-Aware Routing Protocol to Send Video-Reporting Messages Over VANETs in Smart Cities," IEEE Transactions on Vehicular Technology, vol. 66, no. 12, pp. 10611–10625, 2017.

57. S. Kushwaha, S. Shankari, G. Hariharan, K. Vidhya, R. V. K. Reddy and P. Madan, "Kohonen Self-Organizable Maps based Classification of Optical Code Division Multiple Access Codes," 2023 International Conference on Inventive Computation Technologies (ICICT), Lalitpur, Nepal, pp. 1580–1584, 2023.

58. D. Jo and G. J. Kim, "ARIoT: Scalable Augmented Reality Framework for Interacting with Internet of Things Appliances Everywhere," IEEE Transactions on Consumer Electronics, vol. 62, no. 3, pp. 334–340, 2016.

59. P. Fraga-Lamas, T. M. Fernández-Caramés, Ó. Blanco-Novoa and M. A. VilarMontesinos, "A Review on Industrial Augmented Reality Systems for the Industry 4.0 Shipyard," IEEE Access, vol. 6, pp. 13358–13375, 2018.

60. L. Zhang, S. Chen, H. Dong and A. El Saddik, "Visualizing Toronto City Data with HoloLens: Using Augmented Reality for a City Model," IEEE Consumer Electronics Magazine, vol. 7, no. 3, pp. 73–80, 2018.

61. S. Kushwaha, "A Survey on Role of Artificial Intelligence," National Conference on recent advancements in Computational Mathematics and Engineering Sciences, at VIT, India, vol. 1, no. 1, pp. 31, 2019.

62. S. Kushwaha, "Feature Extraction with Convolutional Neural Networks – Deep Learning Approach," National Conference on Recent Advancements in Computational Mathematics and Engineering Sciences, at VIT, India, vol. 1, no. 1, pp. 33, 2019.

9 Phase Noise Effect on Millimeter Wave MIMO-UFMC and OFDM Systems

E. Udayakumar and V. Krishnaveni

9.1 INTRODUCTION

Today's wireless cellular networks are advancing quickly to keep up with the many wireless technology advancements. The new fifth-generation (5G) services and Internet-of-Everything (IoE) apps are mostly to blame for this development. The current focus of 5G research is on resolving filtered orthogonal frequency division multiplexing (OFDM)-based spectral loss and high peak-to-average power ratios are problems with long term evolution-peak to average power ratio (LTE-PAPR). The mm-wave communication is anticipated to power the 5G networks and provide users with access to several gigabits per second. The purpose of the communication is also very high. The precoding techniques can be utilized to boost the spatial multiplexing gain during the radio frequency (RF) stage. For mm-wave communication systems, understanding the problems of hardware faults at the RF stage is critical since they might impair the delivered signals and overall system performance [1].

The success of wireless communication is mirrored by the technology's quick evolution. Different developing technologies, including massive multiple-input multiple-output (MIMO) networks, the Internet of Things (IoT), small-cell networks, and cognitive radio networks are described in a general 5G cellular network design. Millimeter-wave (mm-wave) technology is a significant component of the 5G network and is anticipated to offer hundreds of MHz of bandwidth for wireless transmissions [2]. As a result, small-cell cellular networks are made denser. The approach doesn't necessitate in-depth knowledge of phase noise statistics but may be able to successfully reduce phase noise within each universal filtered multi carrier (UFMC) symbol.

It is predicted that IoT, direct user-to-user communication (D2D), and smart cities would increase exponentially in the next years to connect 20–25 billion devices by the year 2023. There will therefore be a severe spectrum shortage in 5G networks. Massive MIMO techniques, optical wireless networks, and other cutting-edge technologies have all been implemented to boost spectrum usage while enabling very high data speeds [3]. A novel communication paradigm called D2D was first

DOI: 10.1201/9781003407836-9

suggested for cellular networks in order to increase spectral efficiency and decrease latency. D2D communication eliminates the need for BS or the core network. The involvement of the cellular network in the control plane is the primary distinction between D2D and MANET (mobile ad hoc network). In contrast, D2D nodes might take independent action if the cellular infrastructure is disrupted.

Optical wireless communication (OWC) is the most sought-after technology in the communications sector because it can meet the rising demand for capacity. Outdoor communication is the purpose of terrestrial OWC links, also known as FSO communication. Using an LED or laser device, data is communicated via a wireless medium in the form of a light beam. Since it works in an uncontrolled environment, no spectrum usage fees are necessary. The massive MIMO technology is regarded as a breakthrough in wireless communication. A base station (BS) is outfitted with a larger amount of antennas to serve various users simultaneously in a frequency block [4]. Backhauling for tiny cells and interference control in D2D communication systems are two potential uses for huge MIMO systems. Massive MIMO increases the effectiveness of energy radiation.

9.2 LITERATURE REVIEW

To deliver the issue of channel estimation in UFMC transceivers, and they suggest a "flexible" variant of UFMC that enables the integration of numerous frame cognition with various sub-carrier spacing in a single radio carrier. It is mentioned above that discrete-time single-antenna transceivers operating over a straightforward channel, which are frequently described as discrete-time LTI filters. The usage of numerous antennas, on the other hand, is generally known to be a critical component of present (and future) wireless networks. Multiple antennas are required to give diversity operation gains [5].

In fact, one of the three essential technologies required to attain the desired 1000x output boost in comparison to the present-day generation of wireless scheme is the usage of mm-wave. High carrier frequencies enable the utilization of expansive, underused frequency bands, enabling the seamless delivery of Gbit/s data speeds to mobile customers. Mm-wave carrier frequencies, however, are primarily a short range of 100–200 m technology and necessitate the use of multiple antennas due to their increased path-loss and blockage effects. To counteract the increasing path-loss, large array increases are useful [6].

This turns into a performance concern when excessively high data rates are present. Phase noise has a substantial impact on high order modulations like 256 QAM. Phase noise is frequently simulated using multiplicative noise. We will focus on the first model because the local oscillator (LO) is locked in all conventional 802.11 and LTE implementations. Phase noise can be thought of as having two components: a time-varying, frequency-dependent part and a common phase error (CPE) that is shared by all carriers. This component produces the unwanted and damaging inter-carrier interference (ICI) and is often weaker than the CPE. Using an minimum mean squared error (MMSE) equalization in the frequency domain to balance the additive white gaussian noise (AWGN) and the coloured ICI is a well-liked method of ICI mitigation [7].

In order to boost system capacity and provide reliable transmission, OFDM can successfully counteract the multi-path fading channel. The ICI and ISI, which are caused by nearby sub-carriers and symbols, are prevented with OFDM by utilizing a cyclic prefix (CP) and guard interval between the OFDM signals (ISI). Furthermore, OFDM is not advised for use in cognitive radio and IoE networks due to its huge side lobes [8], difficult power amplifier design, which cause synchronization problems and other disadvantages including data aggregation and a high peak-to-average power ratio (PAPR). This inspired scientists to create and suggest reliable modulation methods that can meet the needs of the developing wireless networks. For the 5G air interface, numerous OFDM options have recently been made available. In order to reduce ICI and produce low out-of-band (OOB) emission, the filter bank multicarrier (FBMC) modulation format, for instance, uses a filter bank to individually filter each sub-carrier.

On the other hand, the longer filter length demanded by FBMC results in more complicated hardware and is therefore not practical for many other uses, such as short-packet transfer. In networks using cognitive radio (CR), UFMC is hence better suited for dynamic spectrum allocation methods [9]. The filter's length is shortened as a result. Moreover, UFMC provides multi-service capabilities, decreased latency, lower side lobe radiation, and support for frequency segmentation. It is suitable for quick communication as well. The UFMC uses a Dolph-Chebyshev filter for each sub-band, with a pass band that allows vital signals to flow through without being lost and a stop band that quickly decreases frequency response to prevent OOB emission. The hardware complexity of this technique is high, and the filter tail extends to the subsequent OFDM symbol.

9.3 SYSTEM DESIGN

Phase noise is a proposed issue for this mm-wave MIMO system. Additionally, compared to non-perfect systems, the proposed system will have optimal hardware for mm-wave, even with amplified thermal noise and residual additive transceiver hardware limitations. The spectral efficiency of the suggested system is given under various scenarios (SEs). According to the findings of the analytical and simulation experiments, the phase noise problem may have a negative effect on SE execution, particularly at higher phase noise levels [10].

The current society's thirst for more bandwidth-based applications is fueling 5G research and pushing the cellular industry to produce more advanced technology. The OFDM technology is currently widely used in the wireless field as well as in many contemporary communication schemes due to its effectiveness. TICI and a greater PAPR are disadvantages. When the channel distribution delay exceeds the cyclic prefix length, ISI, that is the reason why CP, happens in OFDM [11].

5G presents fresh methods to deal with these problems. The sub-bands of this waveform are composed of several sub-carriers. The UFMC technique enhances good spectrum efficiency by removing the cyclic prefix and decreasing out-of-band emission. The UFMC technology removes inter-carrier and inter-symbol interference by filtering blocks of sub-carriers before transmission and reception [12]. The generalized filter bank multi-carrier approach and filtered OFDM can both be used

Symbol Mapping	→	Sub-carrier Mapping	→	QAM Modulation	→	UFMC Transmitter

One-Tap equalization	←	QAM Demodulation	←	One-Tap equalization	←	Channel convolution

Sub-carrier De-Mapping	→	Symbol Demodulation	→	Phase Noise reduction	→	Output signal

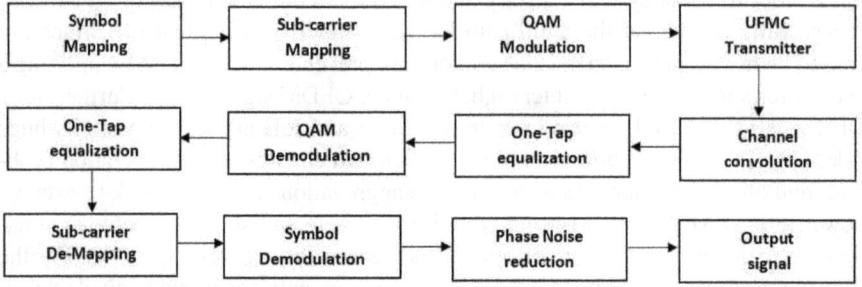

FIGURE 9.1 Block diagram of UFMC systems.

to compare the modulation technique known as UFMC. The base transceiver operation was explained in it. The symbol mapping is to map the symbols of interest information using a QAM modulation techniques. One tap subcarrier bandwidth is less than the channel coherence bandwidth and at the end it detects the phase noise performance. The block diagram of UFMC systems is shown in Figure 9.1.

Therefore, these ICI correction techniques necessitate iterative processing that is more sophisticated and latency-intensive. Another non-iterative maximum likelihood estimation-based technique for reducing frequency is domain phase noise. The work on reducing phase noise discussed above only applies to traditional OFDM systems. Only a little amount of phase noise correction research has been done for UFMC systems. For single-antenna UFMC systems, a non-iterative phase noise mitigation approach was suggested [13]. The method estimates the time domain phase noise and makes adjustments for it using distributed pilots (in the payload). No phase noise mitigation techniques for MIMO-UFMC have yet been suggested. In this study, we expand the single-antenna UFMC system's phase noise mitigation strategy to MIMO-UFMC systems. Mutual coupling, on the other hand, reduces the overall embedded antenna efficiency. The reciprocal coupling effect was formerly described in terms of diversity gain or MIMO capacity. Every sub-band modulated signal is subjected to an L-length FIR filtering. B filtered sub-band modulated signals are added together to create the UFMC signal. The base-band UFMC signal [14] can be

$$Y_{UFMC}(k) = \sum_{i=1}^{B} \sum_{l=0}^{L-1} \sum_{m=0}^{M-1} c_m^i \, e^{\frac{j2\pi(k-1)m}{M}} \, f_i(l), \qquad (9.1)$$

with m the complex-valued symbols for sub-carrier M and sub-band I for each resulting block of length L+M1.

UFMC modulation has drawn a lot of interest among these. It stands for the middle ground between filtered-OFDM and FBMC. The need for sub-band-wise filtering arises from the fact that blocks of sub-carriers frequently experience time-frequency misalignment. Additionally, and perhaps most importantly, the use of filters with a wider bandwidth and a shorter impulse response than those used in FBMC is made possible. Referencing reduces spectral leakage into adjacent sub-bands by applying

Bohman filter-based pulse shaping to the edge sub-carriers of the sub-bands using a combination of antipodal symbol pairs. In order to calculate the weighting factors for interference removal and prevent interference between signals in adjacent channels, the works propose active interference cancellation methods [15].

Then, by fitting a few basis vectors with LS, the dominating phase noise components are calculated. The method reduces to the use of a single basis vector. Phase noise is often not able to be succinctly expressed on a rigid basis, which is the method's fundamental flaw [16]. Massive MIMO communication systems have high complexity because the number of pilot sub-carriers will be greater than the amount of transmitting antennas. We also consider statistical attributes of the PN procedure that vary over time [17]. For practical applications, real-time basis selection is greatly desired since, even when the LO is secured, the statistical characteristics of PN are temporally changing due to environmental factors like temperature, such as heat from a mobile device.

9.3.1 MIMO-UFMC

"Universal filtered multi-carrier" is a kind of modulation method used in wireless schemes is what the abbreviation "UFMC" refers to. OFDM is compared to the lower-frequency radio waves that are generally utilized for wireless communication; mm- waves have distinctive propagation properties. The use of millimeter waves in wireless communication networks has been discussed, and one potential answer is the modulation technique known as UFMC. The effectiveness and dependability of the system may increase as a result. The efficacy of UFMC, one of several modulation schemes being evaluated for use in mm-wave communication systems, will rely on a number of variables, including the particular system needs and the communication channel's characteristics [18].

Mm-wave MIMO systems can be developed utilizing UFMC modulation technology. Using radio frequencies between 30–300 GHz for wireless communication is known as mm-wave communication. By utilizing the channel's spatial diversity, MIMO application makes use of many antennas at the transmitter and receiver to enhance communication performance. In comparison to conventional OFDM, UFMC is a filter bank-based multi-carrier modulation technology that has various advantages, including better spectrum confinement and more resistance against frequency-selective fading. Several antennas are employed at the transmitter and receiver in MIMO-UFMC systems to create numerous parallel data streams [19].

The resulting signals are simultaneously sent over the same frequency range after each stream is UFMC modulated. Advanced signal processing methods are used at the receiver to integrate the signals from all antennas, separate the various data streams, and recover the sent data. The ability of MIMO-UFMC systems to get beyond the constrained bandwidth and significant path loss associated with mm-wave communication is one of their key advantages. Because of their high carrier frequency [20], mm-wave signals are vulnerable to significant route loss and dispersion, which can seriously impair performance. By utilizing the spatial diversity of the channel and cutting-edge signal processing techniques to make up for the channel limitations, MIMO-UFMC systems might lessen these effects.

For high-capacity, high-speed, and dependable wireless communication in mm-wave bands, MIMO-UFMC systems are a promising technology overall. Compared to conventional OFDM-based systems, which are vulnerable to spectral leakage and inter-carrier interference in the high-frequency bands, they provide a more effective and reliable solution [21].

The received time domain wave of the MIMO-UFMC system in a multi-path channel [5] can be

$$Y(n) = \sum_{l=0}^{L-1} H_k \, x(n-l) + x(n), \tag{9.2}$$

where H_k (l = 0,..., L 1) is a $M_t \times M_r$ MIMO channel impulse response (CIR) matrix, x(n) and y(n) are $M_t \times M_r$ transmit and receive signal vectors, and x(n) is a vector made up of the additive white Gaussian noises (AWGNs) at the receive antennas. The internal thermal noise from the RF circuit exceeds the ambient noise in cellular communications. (In most cases, AWGN is used to mimic internal thermal noise.) Multiple routes carry signals from the transmitter to the receiver in multi-path situations. Different (scattering) clusters can further categories' various pathways. Each cluster is further subdivided into a number of sub-paths that have about the same latency but distinct (random) phases [22]. The delay varies between various clusters. One channel tap relates to each cluster. Due to the central limit theorem, the (added) signal from one cluster follows a Gaussian distribution assuming there are numerous sub-paths inside one cluster.

The inverse discrete Fourier transform and filtering are used to create the UFMC modulation once the active sub-carriers are initially organized into B blocks of subsequent sub-carriers (at each transmit antenna). Typically, it is taken that each block has the same amount of sub-carriers in order to make implementation as simple as possible (N0). An NV-length prototype filter can then be exponentially modulated to produce the filter bank. The mathematical expression for the UFMC modulation at the jth transmits antenna is given in [23].

The MIMO-UFMC system's transceiver architecture has been explained throughly. The transmit signals are first converted to QAM symbols at the transmitter before being modulated by UFMC modulators. By using digital-to-analogue converters (DACs), mixers, and the carrier frequency produced by the transmit oscillator, the modulated (digital) signals are up transformed into RF, amplified by power amplifiers (PAs), and transmitted out from the transmit antennas [24]. The received signals are first magnify by low noise amplifiers (LNAs) at the receiver before being down-converted to base-band by mixers utilizing the receive oscillator's carrier frequency, and then converted to the digital domain by analogue-to-digital converters (ADCs).

Keep in mind that due to flaws in the system, there will be phase noise in practice. For accurate data detection, the phase noise that randomly rotates the received signals must be rectified [25]. As a result, the digital signals are first demodulated by UFMC demodulates, then detected by the MIMO decoder, and then demapped into the original signals after passing through the phase noise mitigation (PHM) module to remove phase noise. The DACs, ADCs, PAs, LNAs, and filters in the RF chain have been left out for simplicity's sake. The modulated data is given to inverse discrete

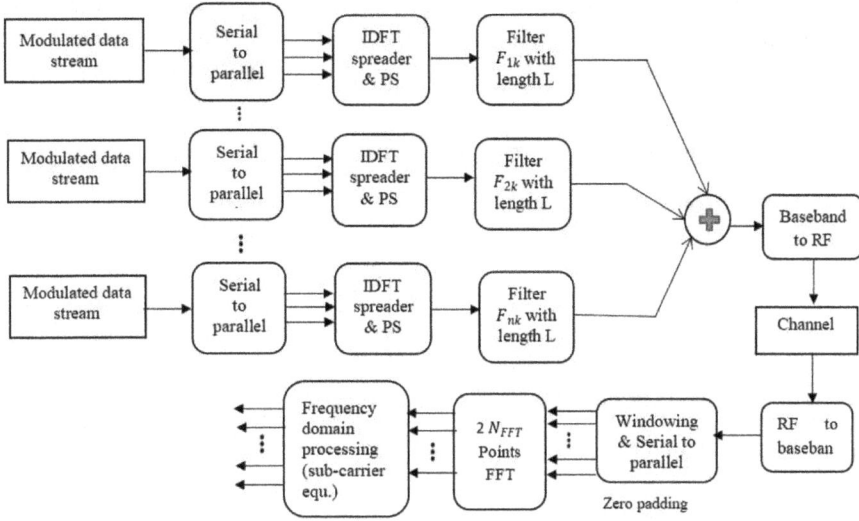

FIGURE 9.2   UFMC transceiver architecture.

Fourier transform (IDFT) to change into frequency domain to digital time samples. Fast fourier transform (FFT) demodulation in the receiver is used. The best method for enhancing the UFMC performance in parallel processing systems is to use FFT for the demodulation. The UFMC transceiver architecture is shown in Figure 9.2.

An effective frequency domain implementation of post-filtering is used to filter a grouping of carriers in the UFMC waveform, which is a derivative of the OFDM waveform [26]. The fact that a resource block serves as the scheduling algorithm's smallest unit in the frequency domain of 3GPP LTE (RB) is the basis for this sub-band filtering operation. In comparison to OFDM, the filtering procedure results in less out-of-band leakage. The B sub-band filtering used in the UFMC transmitter modulates the B data blocks. The shaping filter's temporal transient (tails) causes a spectral efficiency loss even when the transmitted signal does not use CP. The 2 N FFT point FFT that makes up the Rx stage is eliminated by a factor of 2 in order to retrieve the data. Before the FFT, a windowing stage can also be added. Although it causes carrier interference, it is an interesting option to take into account for asynchronous uplink transmissions because it helps to divide immediate users [27].

Unlike the FBMC-UFMC, which sub-carriers to sub-bands before filtering, the number of carriers per sub-band and the filter settings are frequently used. Thus, aliasing is avoided. Non-contiguous sub-bands are still a possibility, allowing for flexible use of the accessible spectrum. As a result, UFMC can be thought of as a compromise between FBMC and OFDM [28].

There is no time connection between the UFMC symbols that are adjacent to one another. N+L-1, where L is the length of filter and N is the FFT size of the IFFT spreaders, gives the duration of a symbol. Like FBMC, UFMC regularly raises the FFT window size, adding to the implementation's complexity. A guard interval may or may not be used as a cyclic prefix in UFMC. This will provide the system more flexibility, improve its resilience to crosstalk, and enable it to take full advantage of the multiple access channel's capabilities (MAC). If N is the number of samples, the amount of complex multiplication operations [7] can be represented as

$$\mu = \frac{N}{2} \, log_2 \, (N).$$

(9.3)

The number of complex addition operations may be formed as

$$\alpha = N \, log_2 \, (N).$$

(9.4)

The bit error rate (BER) is the ratio of the number of bit errors to the total amount of bits transported during the course of the investigated time period. It has no measurement unit and is frequently stated as a percentage:

$$BER = \frac{N_{Err}}{N_{bits}}.$$

(9.5)

An example of a multi-carrier system is the UFMC modulation format, which divides streams with lower data rates in tandem with high data rate signals. For b = 1, 2,..., B, the filtered signals from each sub-band are then combined to produce the time domain UFMC signal. With UFMC, each sub-carrier is filtered in a single step, which has various benefits like lower ICI when there are frequency offsets or jitter. Yet, it makes the signal's computations more complex. Subsets are filtered in the entire band rather than the entire band or a single sub-carrier in UFMC. The Dolph-Chebyshev filter is currently used in the majority of UFMC implementations [29] along with filter optimization methods. It has been found that the Dolph-Chebyshev filter's side-lobe fall rate is flat, increasing the spectrum leakage brought on by out-of-band emissions. Kaiser filter is suggested as an innovative solution to this issue.

9.3.1.1 Phase Noise

In mm-wave communication systems, phase noise, a type of noise that modifies a signal's phase, can be problematic. High-speed communication technologies, such as 5G, are increasingly using mm-wave frequencies from 30–300 GHz. Millimeter wave systems are susceptible to phase noise from a number of sources, including thermal noise, oscillator noise, and ambient noise. Phase noise in high-speed communication systems can result in issues including data transmission mistakes, deteriorated signal quality, and decreased system capacity. Using top-notch oscillators with low phase noise properties is one technique to reduce phase noise in mm-wave systems [19].

9.3.1.2 Iterative Frequency Domain Equalization (IFDE)

In mm-wave communication systems, a method known as iterative frequency domain equalization (IFDE) is employed to mitigate the significant channel defects that exist in this high-frequency range. The enormous available bandwidth in mm-wave systems allows for extremely rapid data transmission [30]. Yet when they go through the environment, the signals at these frequencies suffer from considerable attenuation, reflection, and scattering. As a result, the channel experiences high delay spread and frequency selectivity, which can result in ISI and ICI. A multi-stage equalization method that operates in the frequency domain is called IFDE. The steps involved are as follows:

Channel estimation: Based on the pilot symbols it has received, the receiver calculates the frequency domain channel response.

Equalization: To reduce channel distortion, the receiver employs a linear equalizer in the frequency domain. However, a single equalization solution might not be adequate because the channel response can vary greatly across the bandwidth.

Estimation of residual interference: The residual interference, or the discrepancy between the received signal and the equalized signal, is then calculated using the equalized symbols.

Suppression of interference: A non-linear suppression function is then used to suppress the remaining interference. The interference brought on by noise, neighboring channel interference, and nonlinear distortion is lessened as a result of this step.

Iteration: Up until the residual interference reaches a predetermined level, steps 2-4 are performed repeatedly.

By minimizing the impact of channel distortion and interference, IFDE can greatly enhance the performance of mm-wave communication systems. Unfortunately, it can result in higher latency and demands a lot of processing power. In situations where large data rates are necessary and extremely changeable channel conditions are present, it is therefore generally used [16].

9.3.1.3 Kaiser Filter

In order to reduce phase noise, mm-wave communication systems frequently employ the digital filter known as the Kaiser filter. When a signal's phase swings erratically due to several factors, including oscillator jitter, thermal noise, and interference, it is called phase noise. Phase noise can considerably reduce the performance of mm-wave systems, especially at higher frequencies where the noise power is rather considerable [17].

A finite impulse response (FIR) filter known as a Kaiser filter is one that was created using the Kaiser window function. The filter's transition bandwidth and stopband attenuation are managed using a mathematical formula called the Kaiser window function. In mm-wave systems, where the signal bandwidth is often very broad, the window function offers a technique to balance the filter's passband ripple and stopband attenuation.

In mm-wave systems, the Kaiser filter has a number of benefits for lowering phase noise, including:

Better filtering performance: As compared to other filters, the Kaiser filter offers a sharper transition from the passband to the stopband. This indicates that it can offer higher noise attenuation for components of the signal that are outside of its bandwidth.

Flexible filter design: The Kaiser filter enables flexible filter characteristic design to satisfy certain system requirements. To improve the efficiency of the system, the filter's passband ripple, stopband attenuation, and cutoff frequency can be changed.

Reduced complexity: The Kaiser filter has a finite number of taps since it is a finite impulse response (FIR) filter. Because of this, it has a lower computational complexity than other types of filters, which is crucial in mm-wave systems where high processing power can be difficult to get.

In conclusion, the Kaiser filter is a useful tool for mm-wave communication systems to lower phase noise. Several mm-wave applications like it because of its capacity to offer sharper filtering performance, flexibility in filter design, and decreased complexity.

9.4 SIMULATION AND DISCUSSION

In this section, we run simulations to look into the impacts of phase noises on oscillators that are both free-running and phase locked loop (PLL)-based. We use the assumption that there are 1,024 sub-carriers total in both scenarios, of which 832 are data sub-carriers, 32 are distributed pilots, and 160 are guard band sub-carriers. A total of 64 quadrature amplitude modulation (64 QAM) symbols are put into the active sub-carriers. For the purpose of reducing phase noise, $q = 7$ unknown (see section 9.3) are chosen. The prototype filter for the UFMC modulation is the 64-tap Dolph-Chebyshev filter with 40 dB stop-band suppression. It is anticipated that the channel will have three taps and an equal average tap gain of 1/3. After 20 UFMC symbols, assuming the channel doesn't change, a realization of an independent channel is drawn. In frequency-selective fading channels, UFMC and F-OFDM systems can both outperform conventional OFDM in terms of BER performance. F-OFDM has a higher spectral efficiency whereas UFMC has a superior spectral confinement property and a lower PAPR. The specific application needs and system limitations determine which of these two systems should be used. The BER performance for UFMC and F-OFDM systems is shown in Figure 9.3.

The key distinctions and technical difficulties between using MIMO transceivers at mm-wave carrier frequencies and conventional sub-6 GHz frequencies are shown. Hardware limitations prevent the use of fully digital beamforming structures, which is particularly true at mm-wave frequencies. Additionally, phase noise, which primarily results from flaws in the local oscillator in the transceivers, negatively affects system performance and cannot be disregarded [16]. As a result, there are distinct requirements and limits for wireless transceiver design for mm-wave carrier frequencies than there are for systems operating at sub-6 GHz frequencies. Because they employ various modulation techniques, the signal power vs. time for F-OFDM and UFMC will seem different. Both F-OFDM and UFMC's signal power vs. time plots will display a succession of rectangular pulses, but because each technique uses a different filtering and modulation scheme, the spacing and power of the pulses will vary. The simulation of signal power analysis for UFMC and F-OFDM systems is shown in Figure 9.4.

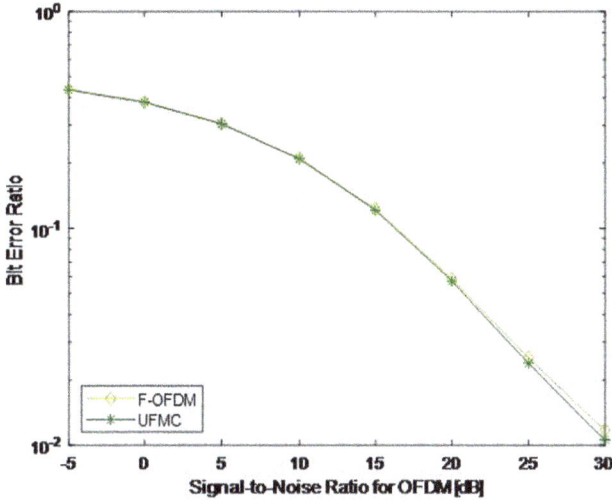

FIGURE 9.3   BER performance for UFMC and F-OFDM systems.

However, offsets for phase noise in OFDM schemes have been thoroughly researched in the literature. For instance, by modulating one symbol to two neighboring sub-carriers with alternate weights, an ICI self-cancellation technique was developed. The SE is reduced fractionally, but the phase noise impact can be efficiently mitigated. A proposed method for frequency domain ICI correction estimates

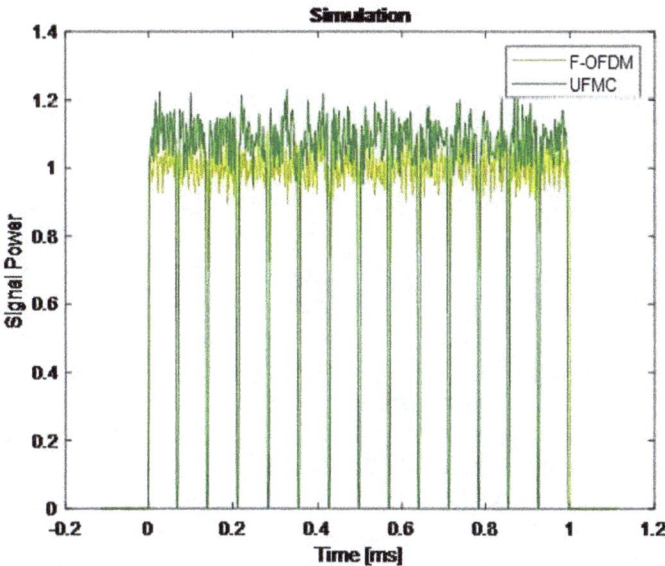

FIGURE 9.4   Simulation of signal power analysis for UFMC and F-OFDM systems.

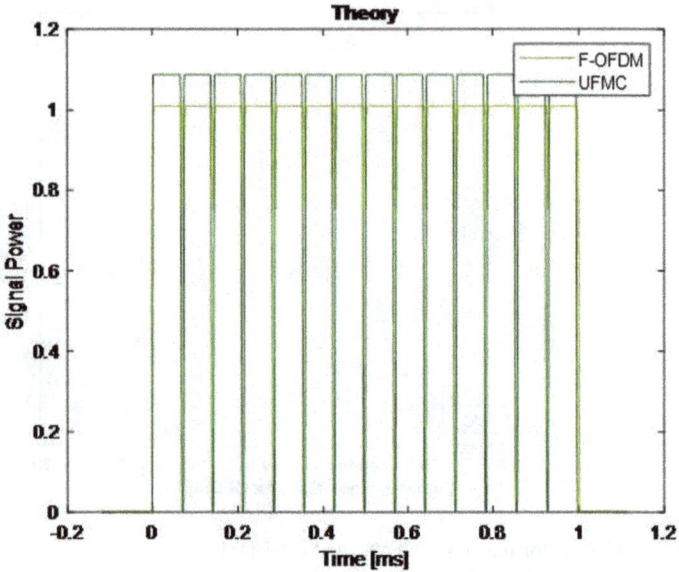

FIGURE 9.5 Signal power and time analysis for UFMC and F-OFDM systems.

the discrete spectral components of phase noise in order to remove the phase noise effect. By adding linear interpolation between neighboring OFDM symbols, the technique was further improved, which improved system performance. In contrast, F-OFDM still needs a cyclic prefix to reduce ISI, which lowers the amount of time-domain resources that are accessible. To cut down on the amount of time that is lost, the length of the cyclic prefix can be tuned based on the channel characteristics. In terms of signal power and time analysis, UFMC and F-OFDM each have their own advantages and disadvantages. Which technique is best suited for particular use cases and applications will require more investigation. The signal power and time analysis for UFMC and F-OFDM systems is shown in Figure 9.5.

In this study, we make use of a NOMA-MIMO system's performance for both ideal and realistic schemes. Also, it may be said that the OFDM system is inferior than the standard UFMC with Dolph-Chebyshev and UFMC (Kaiser), which has identical PAPR characteristics. Even in loud environments, the UFMC system exhibits superior performance to the Kaiser-Bessel window. Further research on UFMC with various multi-path fading channel 5G scenarios is possible. Several aspects of UFMC and F-OFDM systems, such as the usage of filtering and the quantity of sub-carriers, have an impact on their power spectral density (PSD). As a result, there may be a big difference between the two systems' simulated PSDs. The simulation of PSD for UFMC and F-OFDM systems is shown in Figure 9.6.

The frequency-selective multi-path channel is effectively handled by the multi-carrier technique by splitting it up into numerous frequency-flat sub-channels. Numerous multi-carrier methods have recently proposed. One of these is the UFMC, sometimes known as the universal filtered OFDM (UFOFDM). Using

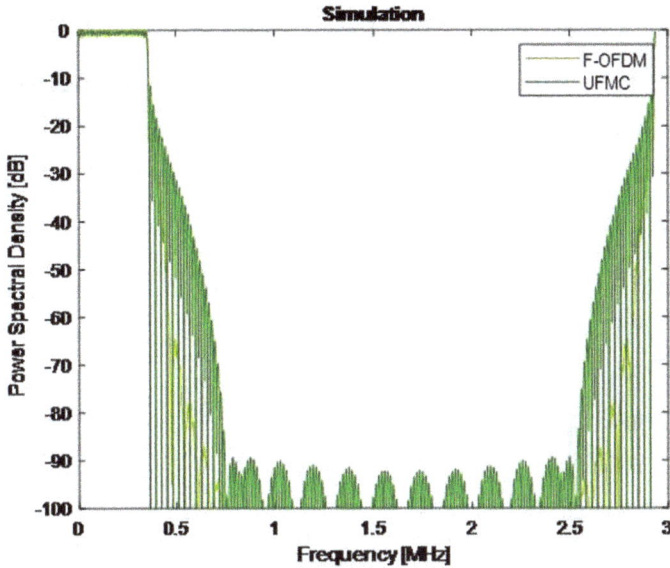

FIGURE 9.6 Simulation of PSD for UFMC and F-OFDM systems.

the simple one-tap channel equalization at each sub-carrier will result in high throughput. The UFMC sub-carrier spacing must be sufficiently narrow, similar to CP-OFDM (or just OFDM). High modulation order is also preferred. However, the oscillator phase noise causes ICI and CPE with denser sub-carriers or higher modulation orders, which can seriously impair performance. The distribution of signal power as a function of time delay is described by the power delay profile (PDP), which characterizes the propagation channel in a wireless communication system. Every wireless communication system, including UFMC systems, depend on the PDP to function properly. A key factor that impacts the performance of a UFMC system is the PDP, which can be computed using channel models that take into consideration the system's features. The power delay profile of UFMC systems is shown in Figure 9.7.

As a result, it is important to get statistical understanding of the PN covariance matrix. Many communication technologies, including WLAN, mm-wave systems, and full-duplex systems, exhibit phase noise. Phase noise is particularly detrimental to full-duplex systems since self-interference is used to separate the conventional signals, as can be seen in the example. Phase noise makes it difficult for transmitters to correctly subtract their own interference, which is detrimental in high order modulations that require high received SNR values that translate to low error vector magnitudes (EVM). The phase noise has an impact on OFDM mm-wave systems as well. The influence of phase noise is almost pronounced when using full order modulations. Coherent optical OFDM-based optical communication systems also exhibit phase noise. Without the use of an optical phase-locked loop, the phase is calculated digitally in these networks, but phase noise causes the estimation to be

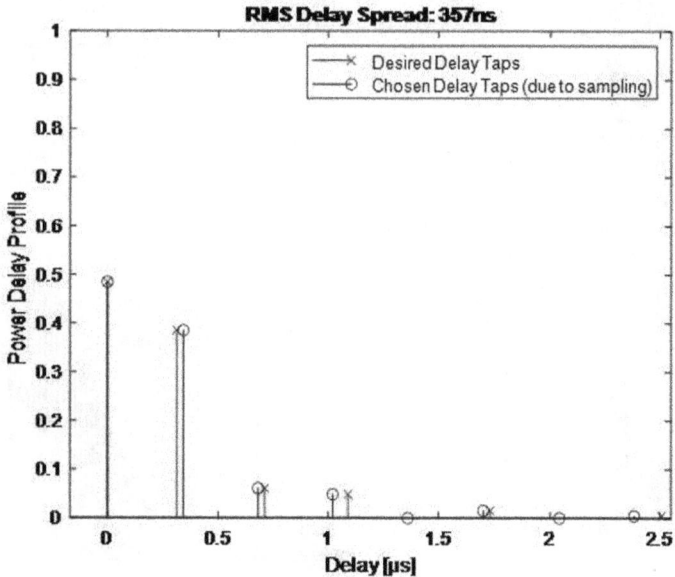

FIGURE 9.7   Power delay profile of UFMC systems.

inaccurate. All of these systems can transmit OFDM signals using the phase compensation approach that we provide.

9.5   CONCLUSION

This chapter surveys phase noise and antenna mutual coupling on MIMO-UFMC systems. The former is assumed to be at sub-6 GHz frequency, whereas the latter is assumed to be at 28 GHz frequency. The latter may have smaller CPE effect (due to the PLL), yet much severer ICI effect than the former. An effective phase noise decrease strategy was conferred. It was shown that phase noise is reduced using CPE correction regardless of the mutual coupling effect. For example, the SER can be reduced from 0.024 to 0.001 at 30 dB SNR, when the mutual coupling effect is taken into account, the phase noise reduction strategy is superior to the CPE correction for PLL-based oscillators. The effectiveness and dependability of the system may increase as a result. The phase noise effect on MIMO-UFMC is reduced using iterative frequency domain equalization (IFDE) by using Kaiser filter. The efficiency of UFMC, one of several modulation schemes being evaluated for use in mm-wave communication systems, will rely on a number of variables, including the particular system needs and the communication channel's characteristics.

REFERENCES

1. Y. Tao, L. Liu, S. Liu and Z. Zhang, "A survey: Several technologies of non-orthogonal transmission for 5G," China Communications, vol. 12, pp. 1–15, 2015.

2. P.N. Rani and C. S. Rani, "UFMC: The 5G modulation technique," Proceedings of the 2016 IEEE International Conference on Computational Intelligence and Computing Research (ICCIC), Chennai, India, 15–17 December 2016; pp. 1–3.

3. F. Schaich and T. Wild, "Waveform contenders for 5G - OFDM vs. FBMC vs. UFMC," 2014 6th International Symposium on Communications, Control and Signal Processing (ISCCSP), Athens, Greece, 2014, pp. 457–460.

4. G. Hussain and L. Audah, "BCH codes in UFMC: A new contender candidate for 5G communication systems," Bulletin of Electrical Engineering and Informatics, vol. 10, no. 2, 904–910, 2021.

5. X. Chen, S. Zhang and A. Zhang, "On MIMO-UFMC in the presence of phase noise and antenna mutual coupling," Radio Science, vol. 52, pp. 1386–1394, 2017.

6. E. Udayakumar and V. Krishnaveni, "Analysis of various Interference in Millimeter-Wave Communication Systems: A Survey," Proceedings of 10th IEEE International Conference on Computing, Communication and Networking Technologies (ICCCNT 2019), Indian Institute of Technology Kanpur, Uttar Pradesh, 2019, pp. 1–5.

7. F. S. Shawqi, et al., "A new SLM-UFMC model for universal filtered multi-carrier to reduce cubic metric and peak to average power ratio in 5G technology," Symmetry, vol. 12, p. 909, 2020.

8. S. A. Fathy, M. Ibrahim, S. El-Agooz and H. El-Hennawy, "Low-complexity SLM PAPR reduction approach for UFMC systems," IEEE Access, vol. 8, pp. 68021–68029, 2020.

9. E. Udayakumar and V. Krishnaveni, "Analysis of Phase Noise issues in Millimeter wave systems for 5G communications," Wireless Personal Communications, Springer, vol. 126, pp. 1601–1619, July 2022.

10. A. Leshem and M. Yemini, "Phase noise compensation for OFDM systems," IEEE Transactions on Signal Processing, vol. 65, no. 21, pp. 5675–5686, Nov. 1, 2017.

11. N Anughna and M Ramesha, "Performance analysis on 5G waveform candidates for MIMO technologies," 2022 IEEE 2nd Mysore Sub Section International Conference (MysuruCon), Mysuru, India, 2022, pp. 1–8.

12. O. S. Faragallah, H. S. El-Sayed and M. G. El-Mashed, "Estimation and tracking for millimeter wave MIMO systems under phase noise problem," IEEE Access, vol. 8, pp. 228009–228023, 2020.

13. V. Vakilian, T. Wild, F. Schaich, S. Ten Brink and J. F. Frigon, "Universal-filtered multi-carrier technique for wireless systems beyond LTE," Proceedings of the 2013 IEEE Globecom Workshops (GC Wkshps), Atlanta, GA, USA, 9–13 December 2013, pp. 223–228.

14. A. A. Ghaith, "Universal-filtered multi-carrier: A waveform candidate for 5G," International Journal for Innovation Education and Research, vol. 6, no. 6, pp. 37–50, 2018.

15. R. T. Kamurthi, "Review of UFMC technique in 5G," Proceedings of the 2018 International Conference on Intelligent Circuits and Systems (ICICS), India, 19–20 April 2018, pp. 115–120.

16. Y. I. Hammadi, R. K. Ahmed, O. A. Mahmood, A. Muthanna, "New Filtering Method to Reduce PAPR and OOBE of UFMC in 5G Communication System," In: Vishnevskiy, V.M., Samouylov, K.E., Kozyrev, D.V. (eds) Distributed Computer and Communication Networks: Control, Computation, Communications. DCCN 2022. Lecture Notes in Computer Science, vol 13766. Springer, Cham, 2022.

17. R. S. Yarrabothu and U. R. Nelakuditi, "Optimization of out-of-band emission using kaiser-Bessel filter for UFMC in 5G cellular communications," China Communications, vol. 16, no. 8, pp. 15–23, Aug. 2019, doi: 10.23919/JCC.2019.08.002.

18. E. A. Tuli, J. M. Lee and D. -S. Kim, "Improved partial transmit sequence based PAPR reduction of UFMC systems," 2022 International Conference on Information Networking (ICOIN), Jeju-Si, Korea, Republic of, 2022, pp. 391–395.

19. S. Sidiq, J. A. Sheikh, F. Mustafa, et al., "A new genetic algorithm bio-inspired based impartial evaluation of UFMC and GFDM under diverse window constraints," Arabian Journal for Science and Engineering, vol. 47, pp. 14173–14184, 2022.

20. R. Wang, H. Mehrpouyan, M. Tao and Y. Hua, "Channel estimation, carrier recovery, and data detection in the presence of phase noise in OFDM relay systems," IEEE Transactions on Wireless Communications, vol. 15, no. 2, pp. 1186–1205, Feb 2016.

21. L. Dai, Z. Wang and Z. Yang, "Time-frequency training OFDM with high spectral efficiency and reliable performance in high speed environment," IEEE Journal on Selected Areas in Communications, vol. 30, pp. 695–707, 2012.

22. F. Schaich and T. Wild, "Waveform contenders for 5G-OFDM vs. FBMC vs. UFMC," Proceedings of the 2014 6th International Symposium on Communications, Control and Signal Processing (ISCCSP), Athens, Greece, 21–24 May 2014, pp. 457–460.

23. L. Samara, M. Mokhtar, Ö. Özdemir, R. Hamila and T. Khattab, "Residual self-interference analysis for full-duplex OFDM transceivers under phase noise and I/Q imbalance," IEEE Communications Letters, vol. 21, no. 2, pp. 314–317, Feb 2017.

24. E. Udayakumar and V. Krishnaveni, "Phase noise effect on millimeter wave MIMO-OFDM and FBMC-OQAM systems," AIP Conference Proceedings of International conference on Applied Data Science and Smart Systems, Chitkara University, Punjab, 4th & 5th Nov, 2022.

25. E. Udayakumar and V. Krishnaveni, "Performance evaluation of phase noise reduction for MIMO FBMC-OQAM systems," Solid State Technology, vol. 63, no. 5, pp. 5668–5675, Nov 2020.

26. E. Udayakumar and V. Krishnaveni, "A review on interference management in millimeter-Wave MIMO systems for future 5G networks," Proc. of 4th International Conf. on Innovations in Electrical and Electronics Engineering (ICIEEE 2019), Guru Nanak Institute of Technology, Hyderabad, 2019, pp. 715–721.

27. S. Buzzi, C. D'Andrea, D. Li and S. Feng, "MIMO-UFMC transceiver schemes for millimeter-wave wireless communications," IEEE Transactions on Communications, vol. 67, no. 5, pp. 3323–3336, May 2019.

28. B. O. Ahmed, A. Abdikarin Ali, M. A. Hussein, S. Mohamed Isse, A. M. Hussein and B. Abdirahman Hussein, "A study on the performance metrics of the universal filtered multi carrier waveforms for 5G," 2021 International Conference on Forthcoming Networks and Sustainability in AIoT Era (FoNeS-AIoT), Nicosia, Turkey, 2021, pp. 72–77.

29. F. S. Shawqi, et al., "A new SLM-UFMC model for universal filtered multi-carrier to reduce cubic metric and peak to average power ratio in 5G technology," Symmetry, vol. 12, p. 909, 2020.

30. N. Shaik and P. K. Malik, "Modified entropy based least square channel estimation technique for 5G massive multiple input multiple output universal filtered multicarrier systems," Engineering Research Express, vol. 5, p. 015050, 2023.

10 Role of Metalens in Designing Effective Antennas towards the Urban Transformation in Sub-6 GHz Bands

Solution from the Design Prospective

Bikash Ranjan Behera and
Soumya Ranjan Mahapatro

10.1 INTRODUCTION

Recent times have witnessed a technological revolution in antenna design due to the introduction of 5G/6G communications and with the advent of applications like radio frequency (RF) energy harvesting and wireless power transmission. The prefix "meta" in the term "metamaterials" (MTMs) often refers to something beyond or altered. These materials have unique property; they can change into something that is not found on earth. No material on earth is believed to have negative permeability, and all natural materials on earth have positive permittivity, permeability, and refractive index. Strangely, they are usually responsible for manipulating the electromagnetic waves [1, 2], enthusiastically for enhancing antenna performance [3] in the context of contemporary RF applications.

Research [1–3] shows that high-gain antennas are usually needed in wireless communication systems. Their importance lies in the fact that they improve signal strength by reducing interference and lowering out the free space path loss (FSPL). When judging how well an antenna works, the CP antenna gain should be considered the most important factor. Although MTM strategies have been proposed in [4–14], as a means of enhancing antenna gain [2], the practical aspect for increasing working bandwidth and circular polarization (CP) features like axial ratio bandwidth (ARBW), CP gain, and antenna efficiency, which is still relatively unknown [3] for the desired operating bands.

Utilization of 3.5/5GHz band creates advancements in wireless communication systems that have resulted in the emergence of RF-based urban transformation.

They typically have an analog and digital topology, which calls for high antenna gain, wide bandwidth responses, increased system capacity, and dynamic switching features capabilities in order to achieve CP for effective signal coverage in small base stations. As a result of these advantages, antennas are highly sought after and necessitate technical qualities like low profile, higher gain, and maximized polarization capabilities that can be used as a crucial resource for nearby wireless connectivity [1].

Out of all the various CP-inspired antenna systems that are available, this one has become very common because system integration is becoming more and more important to improving performance in wireless applications [1, 3]. While the progress has been made in the areas of modeling and achieving low-profile features of RF front-ends, these problems still present a significant technical challenge [1] and the new designs (i.e., the incorporation of a feasible solution) are needed to address them, despite the progressive developments in [4–14]. Although CP have been accomplished as a result of geometrical incursions [4–8] and incorporation of MTMs/metasurfaces [9–14], the achievement of limitation as reported in [1–3], which is the fundamental target, has not been yet attained. Metalens, when applied to printed monopole antenna, significantly enhances the CP performance [9–14].

Further, the radiation stability appears to be a concern when PEC reflectors are added for bandwidth and gain augmentation. Due to effective electromagnetic wave reflecting properties, metalens are introduced to make up for the shortcomings observed in the PEC reflectors. In the subsequent section, pertinent insights with their corresponding working methodologies are highlighted.

10.2   ANTENNA DESIGN AND THEIR RELEVANT OUTCOMES

The usefulness of a Y-shaped monopole radiator (YSPMA) depends on one very important thing: tuning the impedance characteristics to get better results in the desired bands (3.5/5 GHz). The input resistance of a monopole antenna is resistive, with 50-Ω at the center resonant frequency, where the antenna length is equal to one-quarter of the wavelength (λ). In [15], the PMA behaves like a capacitor at frequencies below their resonant frequency, but their input resistance is inductive at higher frequencies. It should be kept in mind that the length of a monopole is longer than $\lambda/4$. The geometrical schematic along with its developmental stages are shown in Figures 10.1a and b respectively.

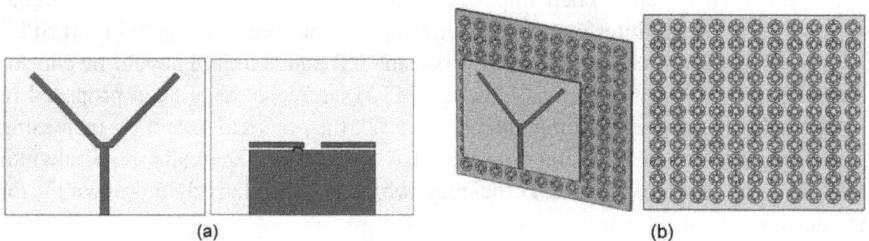

(a)   (b)

FIGURE 10.1   Schematic: (a) Without metalens and (b) with metalens.

TABLE 10.1

An Evaluation of the Antenna's Characteristics with Regard to Previously Developed Antennas [4–14]

References	IBW	ARBW	CP Antenna Gain
[4]	0.62%	0.18%	6.97 dBic
[5]	3.77%	1.86%	6.3 dBic
[6]	6%	3.3%	2.7 dBic
[7]	15.2%	3.2%	4.1 dBic
[8]	16%	10%	5.5 dBic
[9]	22.6%	14.3%	4.2 dBic
[10]	33.7%	16.5%	5.8 dBic
[11]	29.41%	9.05%	6.34 dBic
[12]	16.36%	11.93%	5.62 dBic
[13]	20.5%	15.5%	4.25 dBic
[14]	3.9%	2.2%	1 dBic
Proposed Work	**50.42%**	**16.87%**	**7.53 dBic**

For this reason, printed monopole antennas (PMAs) are frequently employed in UWB bands (i.e., from 3.1–10.6 GHz). However, its deployment in the 3.5/5 GHz frequency is less prevalent [3]. The PMAs have been shown to have significant limitations (frequency dependency) due to their inability to respond in the afore-mentioned frequency band, documented in [4–14] with more and more responsive behavior towards linear polarization (LP) traits. Table 10.1 shows a comparative analysis that illustrates the generic solution presented in Figure 10.1. Their outcomes are shown in Figure 10.2, along with its developmental stages in Figure 10.3. Additionally, Table 10.2 contrasts the capability of PMA with and without the presence of metalens.

10.3 UNDERSTANDING THE POLARIZATION AND METALENS FEATURES: FROM CONCEPT TO ITS CEM ANALYSIS

Here, the proposed metalens-inspired PMA reported for $f = 5$ GHz is concerned with achieving stronger CP attributes. The interpretation of the CP mechanism is shown in Figures 10.4a–d.

CP analysis begins with a first method (i.e., a study of the surface current distribution). At first, the parasitic conducting strip is not present, so there is no connection created with the partial ground plane. Since the induced surface currents appearing on the horizontal borders of the partial ground plane are seen to flow in the opposite direction, they cancel out, only the vertical surface currents on the monopole arm remain. So, a linearly polarized (LP) wave is created. As a result of it, the rearrangement of the surface currents already present on the parasitic conducting strips and the partial ground plane, the currents on the upper edge and the lower edge of the partial ground plane appear to be flowing in the same direction, producing horizontal

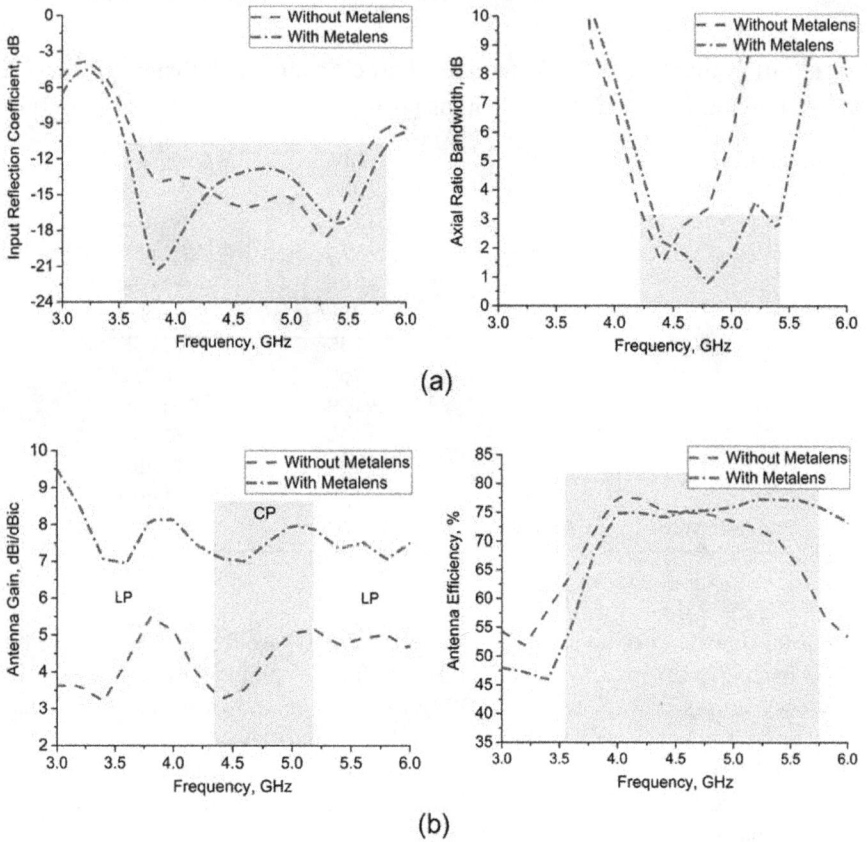

(a)

(b)

FIGURE 10.2   Performances: (A) IBW, ARBW and (b) CP antenna gain, antenna efficiency.

surface currents. As shown in Figure 10.4a, the horizontal and vertical current components act as the main forces behind the development of CP characteristics.

The distribution of electric fields provides a second, more basic method for assessing CP. Here, the CP antenna realizes LHCP, when a parasitic conducting strip is coupled to the partial ground plane. Thus, as illustrated in Figure 10.4b, the electric field vectors rotate clockwise when the phase shifts from 0° to 90°. Existence of a left-handed circular polarized wave (LHCP) is confirmed by their orthogonal properties. So, the third and fourth primitive approaches to CP analysis are through the normalized far-field radiation pattern. It is proportional to the relative power of the radiation pattern. The wideband CP antenna offers exceptional LHCP properties. Figures 10.4c and 10.4d show that LHCP is significantly stronger than RHCP at the 5 GHz.

The fifth approach of analyzing CP traits is presented as mathematical interpretation of gain-bandwidth product, as in equation (10.1):

$$C_{cr} = F(BW_{3-dB}, G_{3-dB}),   \tag{10.1}$$

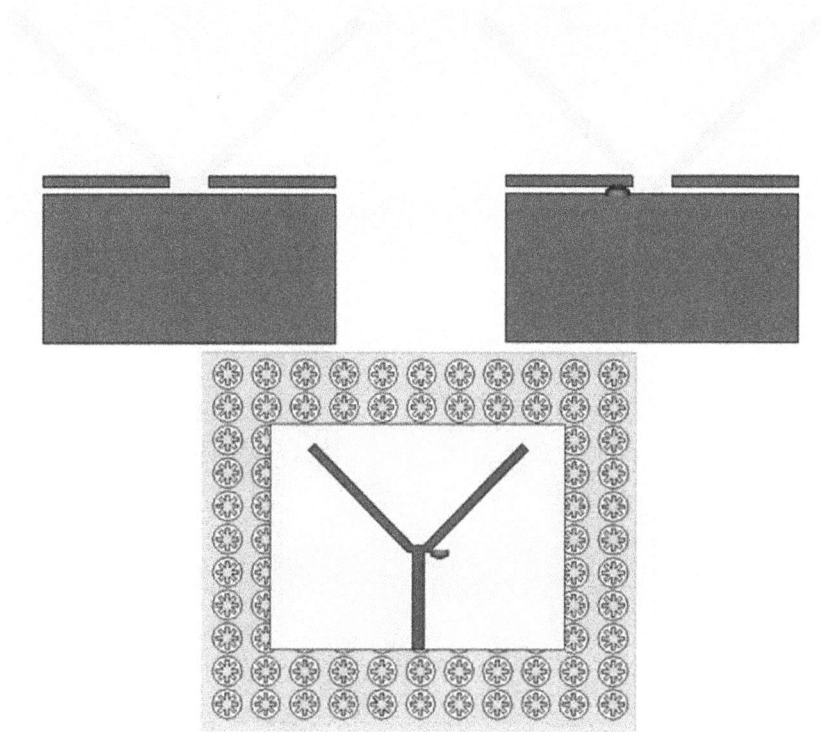

FIGURE 10.3   Developmental stages are: LP antenna without metalens, CP antenna without metalens, and CP antenna with metalens (final stage).

then,

$$C_{cr}^{'} = BW_{3-dB} \times G_{3-dB}^{'},$$ (10.2)

and furthermore,

$$C_{cr} = \frac{BW_{3-dB} \times G_{3-dB}}{100}.$$ (10.3)

TABLE 10.2

A Comparison of Antenna Performances with and without the Presence of Metalens

Antenna Parameters	Without Metalens	With Metalens	Improvement
IBW	45.21%	50.42%	Yes
ARBW	9.23%	16.87%	Yes
CP Antenna Gain	2.98 dBic	7.53 dBic	Yes
Antenna Efficiency	> 70%	> 75%	Yes
Operating Bands	3.5/5 GHz		

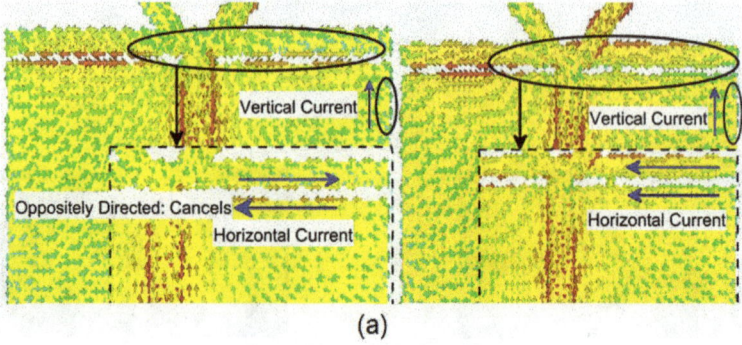

(a)

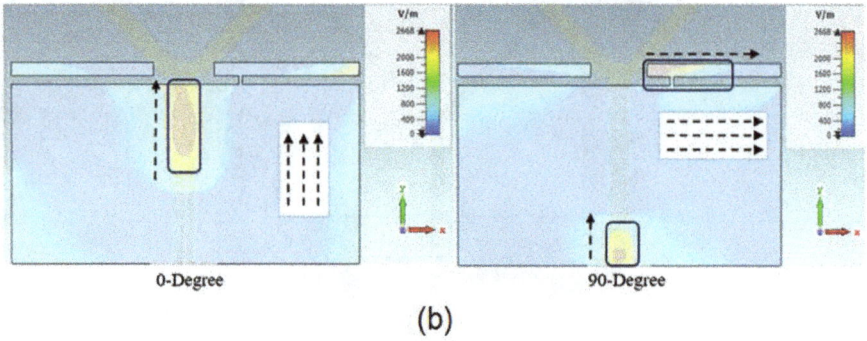

0-Degree   90-Degree

(b)

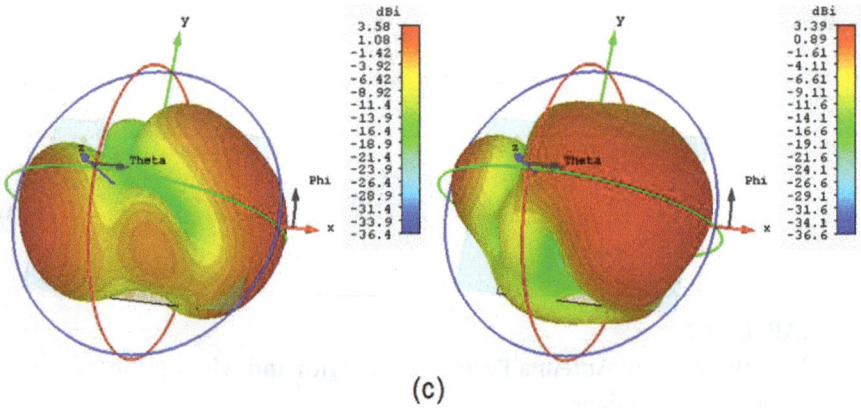

(c)

FIGURE 10.4 CP analysis at 5 GHz: (a) 1st method, (b) 2nd method, (c) 3rd method, and (d) 4th method. *(Continued)*

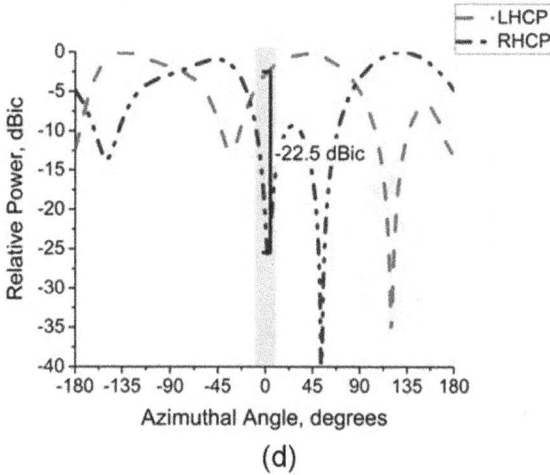

(d)

FIGURE 10.4   *(Continued)*

When compared to other antennas, the CP antenna is virtually identical; however, by using C_{cr-1}-C_{cr-4} we may overcome the restrictions of the conventional methods of comparison, while taking consideration of the CP attributes. Since this is a generalized equation for evaluating the CP antenna, so we can consider equation (10.3). To reinforce the investigation in this context, a comparative study is given in Table 10.3 in relation to the stated designs in [4–14].

The key to getting these types of antenna performances is geometric understanding [15–18]. Further, the improved E-field distribution concentrated the radiation fields for the occurrence of quasi-TM_{30} modes (transformation of TM_{10} to TM_{30} modes), resulting in improved impedance performances (shown in Figure 10.5). As

TABLE 10.3

A Gain-Bandwidth Comparison with Regard to Antenna Designs [4–14]

References	$BW_{3\text{-}dB}$	$G_{3\text{-}dB \, (avg)}$	$G_{3\text{-}dB \, (peak)}$	$C_{cr\text{-}1}$	$C_{cr\text{-}4}$
[4]	0.18%	6.97 dBic	6.97 dBic	0.01	0.01
[5]	1.86%	6.3 dBic	6.3 dBic	0.01	0.01
[6]	3.3%	2.7 dBic	2.7 dBic	0.08	0.08
[7]	3.2%	4.5 dBic	4.5 dBic	0.14	0.14
[8]	10%	5.5 dBic	6.67 dBic	0.55	0.67
[9]	14.3%	4.2 dBic	4.8 dBic	0.61	0.68
[10]	16.5%	5.8 dBic	5.8 dBic	0.95	0.95
[11]	9.05%	6.34 dBic	6.34 dBic	0.57	0.57
[12]	11.93%	5.62 dBic	5.94 dBic	0.67	0.71
[13]	15.5%	4.25 dBic	4.25 dBic	0.65	0.65
[14]	2.2%	7.1 dBic	7.1 dBic	0.15	0.15
Proposed Work	**16.87%**	**7.53 dBic**	**8.21 dBic**	**1.27**	**1.38**

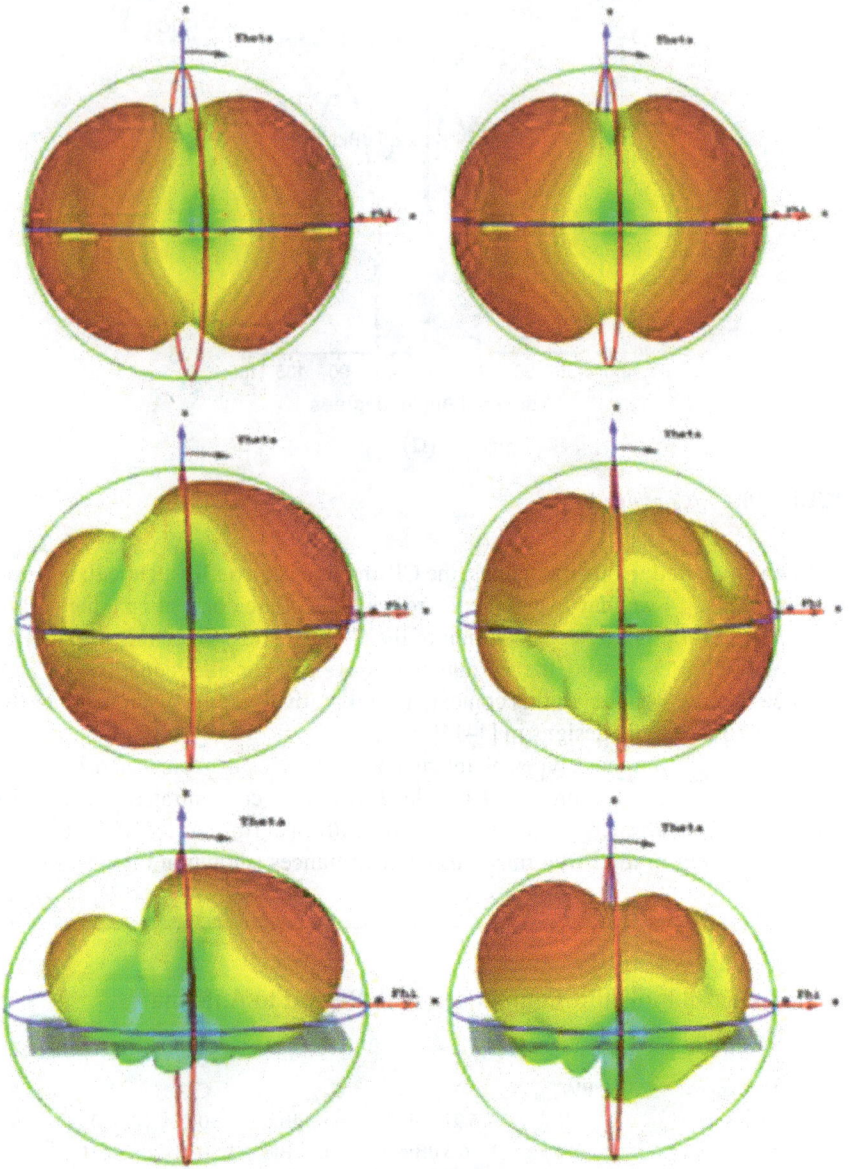

(a)

FIGURE 10.5 Insights: (a) transformation, (b) behavior of electric field, and (c) role of lumped elements (i.e., circuit of metalens). *(Continued)*

E-Field (V/m)

3.0000e+002
2.7857e+002
2.5714e+002
2.3571e+002
2.1429e+002
1.9286e+002
1.7143e+002
1.5000e+002
1.2857e+002
1.0714e+002
8.5714e+001
6.4286e+001
4.2857e+001
2.1429e+001
0.0000e+000

E-Field Coupling Intensity

→ 300 V/m
→ 220 V/m
→ 100 V/m
→ 21 V/m

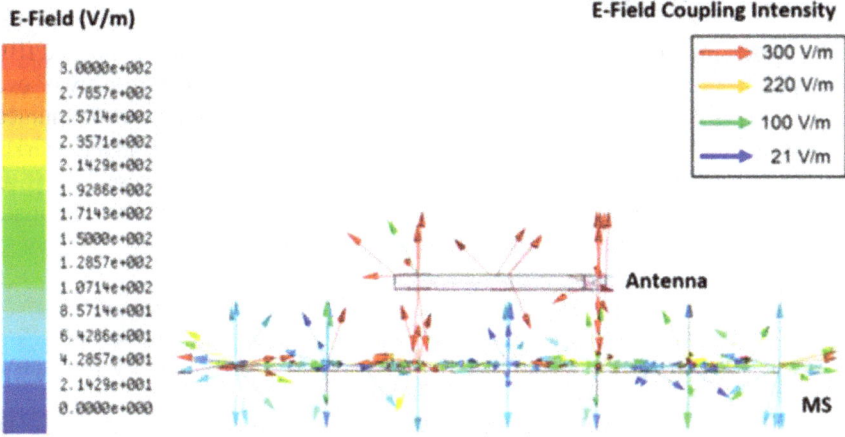

Antenna

MS

(b)

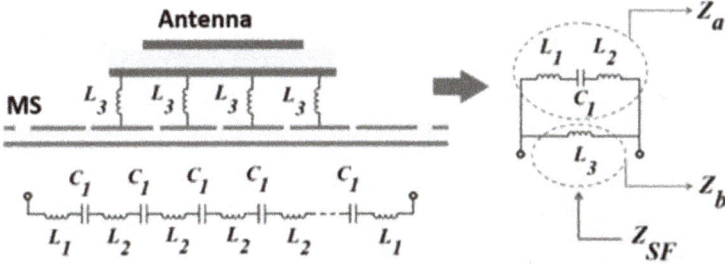

Antenna

MS L_3 L_3 L_3 L_3

C_1 C_1 C_1 C_1 C_1

L_1 L_2 L_2 L_2 L_2 L_1

Z_a

L_1 L_2

C_1

L_3

Z_b

Z_{SF}

(c)

FIGURE 10.5   *(Continued)*

a result of the development of grid-slotted sub-patches (metalens) aid for impedance matching is significant from the standpoint of a device (application perspective).

10.4   CONTRIBUTIONS

a. In the 3.5/5 GHz range, simple monopole antenna and twin parasitic conducting strips are employed with metalens.

b. For optimum radiation efficiency in the $\varphi=0°$ and 90° planes, metalens is loaded at $\lambda/4$ wavelength. It helps in configuring a low-profile design and enhances the antenna connection procedure. As a result of this, radiation is improved in the boresight direction.

c. The gain-bandwidth product is used to study the CP, reported for the very time in the literature [19]. The investigation of CP properties is also highlighted, with the help of four other methods.

10.5 CONCLUSION

This chapter presents a CP metalens-based monopole antenna. It has a widened IBW and ARBW with CP antenna gain of over 7.35 dBic and antenna efficiency of over 70 percent. With improved features, it will play a crucial role in future communication (aimed for RF-based urban transformation), penetrating towards global market, and helping usher in a new age of better connectivity and less power use. Lastly, the chapter provides a motivation to find out general answers for making the metamaterial antennas that can be used in a wide range of situations.

REFERENCES

1. Kumar, P., Ali, T., Pai, M. (2021). Electromagnetic metamaterials: A new paradigm of antenna design. IEEE Access, 9, 18722–18751.
2. Milias, C., Andersen, R., *et al.* (2021). Metamaterial-inspired antennas: A review of the state of the art and future design challenges. IEEE Access, 9, 89846–89865.
3. Behera, B. R., Meher, P. R., Mishra, S. K. (2020). Microwave antennas-An intrinsic part of RF energy harvesting systems: A contingent study about its design methodologies and state-of-art technologies in current scenario. International Journal of RF and Microwave Computer-Aided Engineering, 30, e22148.
4. Ko, S. T., Park, B. C., Lee, J. H. (2013). Dual-band circularly polarized patch antenna with first positive and negative modes. IEEE Antennas and Wireless Propagation Letters, 12, 1165–1168.
5. Cai, T., *et al.* (2015). Low-profile compact circularly-polarized antenna based on fractal metasurface and fractal resonator. IEEE Antennas and Wireless Propagation Letters, 14, 1072–1076.
6. Lin, J. F., Chu, Q. X. (2018). Enhancing bandwidth of CP microstrip antenna by using parasitic patches in annular sector shapes to control electric field components. IEEE Antennas and Wireless Propagation Letters, 17, 924–927.
7. Lam, K., Luk, K., *et al.* (2011). Small circularly polarized U-slot wideband patch antenna. IEEE Antennas and Wireless Propagation Letters, 10, 87–90.
8. Yue, T., *et al.* (2016). Compact, wideband antennas enabled by interdigitated capacitor-loaded metasurfaces. IEEE Transactions on Antennas and Propagation, 64, 1595–1606.
9. Hussain, N., Naqvi, S. I., *et al.* (2020). A metasurface-based wideband bidirectional same sense circularly polarized antenna. International Journal of RF and Microwave Computer-Aided Engineering, 30, e22262.
10. Wu, Z., Li, L., *et al.* (2016). Metasurface superstrate antenna with wideband circular polarization for satellite communication application. IEEE Antennas and Wireless Propagation Letters, 15, 374–377.
11. Rajanna, P., *et al.* (2020). Characteristic mode-based compact circularly polarized metasurface antenna for in-band RCS reduction. International Journal of Microwave and Wireless Technologies, 12, 131–137.
12. Ameen, M., Chaudhary, R. K. (2021). Metamaterial circularly polarized antennas: Integrating an epsilon negative transmission line and single split ring-type resonator. IEEE Antennas and Propagation Magazine, 63, 60–77.
13. Yang, W., Tam, K. W., Choi, W. W., *et al.* (2014). Novel polarization rotation technique based on artificial magnetic conductor and its application in low-profile circular polarization antenna. IEEE Transaction on Antennas and Propagation, 62, 6206–6216.
14. Cao, W., *et al.* (2019). Bandwidth enhanced dual-band patch-coupling microstrip antenna with omnidirectional CP and unidirectional CP characteristics. IET Microwaves, Antennas & Propagation, 13, 584–590.

15. Behera, B. R., *et al.* (2021). Metasurface superstrate inspired printed monopole antenna for RF energy harvesting application. Progress in Electromagnetics Research C, 110, 119–133.
16. Mohanty, A., Behera, B. R. *et al.* (2022). Investigation of dual-layer metasurface-inspired fractal antenna with dual-polarized/-modes for 4G/5G applications. Electronics, 11, 2371.
17. Behera, B. R., Mishra, S. K. (2022). Investigation of a high-gain and broadband circularly polarized monopole antenna for RF energy harvesting application. International Journal of Microwave and Wireless Technologies, First View, vol. 15, 1–12.
18. Mohanty, A., Sahu, S. (2021). Compact wideband hybrid fractal antenna loaded on AMC reflector with enhanced gain for hybrid wireless cellular networks. AEU-International Journal of Electronics and Communication, 138, 153837.
19. Behera, B. R., Mishra, S. K., Alsharif, M. H. Jahid, A. (2023). Reconfigurable antennas for RF energy harvesting application: Current trends, challenges, and solutions from design perspective. Electronics, 12, 2723.

11 Mobile Communications and Networks Beyond 5G

*Abhilasha Gautam, Prabhat Thakur,
and Ghanshyam Singh*

11.1 INTRODUCTION

Similar to previous generation 4G, 5G new radio (NR) seeks to continue the trend of offering a user experience that is exponentially better and offers more diverse applications for cellular communications. Currently, 5G NR outperforms long term evolution (LTE)/LTE-A technology in a number of ways, including higher data rates, low latency, higher mobility (~500km/h), and connectivity to 10^6 devices/km² [1]. The main focus of 5G is to provide the end user a versatile mobile experience by enabling new application. In addition to conventional microwave communication, new technologies including millimeter wave (mm-wave)" and "massive multiple-input multiple-output (m-MIMO) are used to enable the additional services offered by 5G. Error-control, radio, and modulation schemes for these techniques are defined in 5G [2]. Despite 5G's gains, the rollout of 6G networks is inevitable due to the development of smart infrastructure, effective technologies, and a variety of wireless applications. 6G networks are intended to be multi-band, decentralized, fully autonomous, and user-centric systems that enable satellite, aerial, terrestrial, underwater, and underground communications, in contrast to 5G networks. A few examples of technology that will represent today's technological aspirations as tomorrow's reality include massive unmanned mobility in three dimensions, holographic telepresence, eHealth and wellness applications, all-pervasive networking, industry 4.0, "augmented reality (AR)," and "virtual reality (VR)[3]. In order to achieve the 6G's performance indicators and application scenarios, it is projected that the current 5G wireless communication networks will go through four significant paradigm shifts, namely, comprehensive applications, endogenous network security, global coverage, and all spectrums.

This chapter provides some discussions about topics covering 5G/6G usage scenarios, implementation blockheads due to unresolved issues carried from previous generations, and technological trends for beyond 5G networks simultaneously highlighting trends for further research given as shown in Figure 11.1.

11.1.1 5G/6G SERVICES, USE CASES AND KPIs

A comparison between 5G and 6G networks is shown in Table 11.1.

 DOI: 10.1201/9781003407836-11

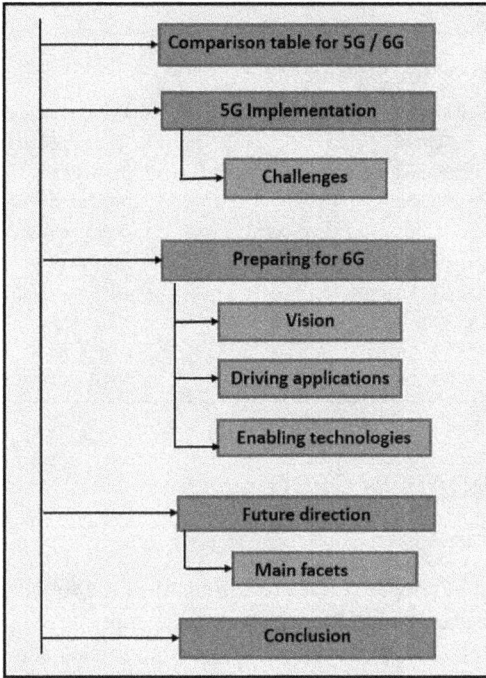

FIGURE 11.1 Layout of the chapter.

TABLE 11.1
Comparative Analysis for 5G and 6G

	5G	6G
Services	• eMBB – "enhanced mobile broadband"	• FeMBB – "further enhanced mobile broad band"
	• URLLC – "ultra reliable low latency communication"	• eURLLC – "extreme ultra reliable low latency communication"
	• mMTC – "massive machine type communication"	• umMTC – "ultra massive machine type communication"
	• Hybrid (URLLC + eMBB) [4]	• LDHM – "long-distance and high-mobility communications"
		• ELPC – "extremely low-power communications"
Use cases [5]	• Virtual Reality/AR	• Teleportation
	• 360° Videos	• Haptic communication
	• Ultrahigh definition video	• Sensory Sensing/Reality
	• Vehicle to everything	• Vehicular Automation
	• IoT	• Industry Automation
	• Smartcity/Factory/Home	• Space Travel
	• Telemedicine	• Deep-sea imaging
	• Wearable Devices	• Internet of Nano-things
		• eHealth

(Continued)

TABLE 11.1 *(Continued)*

Comparative Analysis for 5G and 6G

	5G	6G
Network design	• 5G-NR	• AI-enabled network
	• mm-wave small cell network	• 3D network
	• Slicing	• mm-wave/THz network
		• Cell free mMIMO
Area throughput	• 10 Mb/s/m^2	• 1Gb/s/m^2
Connectivity	• 10^6 devices/km^2	• 10^7 devices/km^2
Latency	• 1 ms	• 10–100 μs
Mobility	• 500 km/h	• ≥1,000 km/h
Spectral efficiency	• 3 × that of 4G	• 5–10 × that of 5G

11.2 IMPLEMENTATION OF 5G

11.2.1 CHALLENGES

Although 5G rollout has begun, there are still many of the sophisticated features that are anticipated to be accessible. Different "verticals" (specifications for various applications) have been created, and some of their requirements contradict one another. In addition to having more needs, 5G also depends on technologies that have not yet been previously used. Extremely effective hardware and software are required for massive MIMO and mm-wave communications. The change has a substantial impact on the core network since there is a large complexity gap between 4G and 5G. It will merely require time and regular effort to get through some of the present difficulties, such as spectrum clearance and auctioning, equipment deployment, and so on. In addition to having the fastest time to market ever, 5G will undoubtedly need numerous tweaks and modifications to get closer to ideal performance [6].

As a result, engineers and researchers will have to deal with the introduction of complicated and novel technologies in the public and industrial sectors at an unprecedented pace and scale in the upcoming years. Figure 11.2, highlights the bottleneck areas of 5G implementation.

Many of the issues that 5G is currently facing have not been resolved yet. Here are a few crucial components that 5G must have in order to succeed fully.

11.2.2 NETWORK ARCHITECTURE

Due to the diversity of applications, it is particularly challenging to assess the performance of a 5G network and change its operating mode. The new network design should be researched and used in accordance with network characteristics and standards to more fully understand what 5G wireless networks would need. Existing models, especially those based on random variable outcomes, either depart from deployment realities by simply allowing for a macro-performance evaluation without allowing for fine-grained network reconfiguration, or they exhibit complexity that we can no longer manage analytically. Improved resource management and

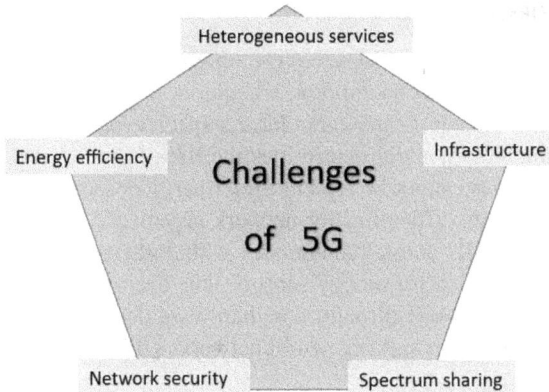

FIGURE 11.2 Challenges of 5G.

orchestration are also necessary to handle the many distinct applications, each of which demands a very different level of performance. Fundamental methods for assessing network performance in a multidimensional space (throughput, latency, energy, dependability, etc.) and tools (ML, AI) are still required for managing 5G more successfully [7].

11.2.3 CHANNEL STATE INFORMATION

Numerous physical layer methods rely on CSI information. However, when taking into account enormous MIMO or IoT, particularly in dynamic situations, it becomes impractical to predict the state for all of the relevant channels. To help the system estimate the CSI, location-aware channel databases or deterministic channel models developed using the actual geometry of the scenario under consideration can be used; however, their real-time use in resource allocation, signal processing, and localization techniques is still in its infancy. Techniques for machine learning may also be very beneficial [8].

11.2.4 MOBILITY LINKS

Beam shaping and massive multiple antenna systems are two further innovative technologies employed in 5G technology. This can be extremely important in the mm-wave regions to combat the significant isotropic free-space path loss we incur while shrinking the wavelength. It is crucial to build efficient beam guiding and tracking. High-speed data transfer would be made possible in the context of smart mobility by providing radio communication in the mm-wave band between vehicles and infrastructure or among vehicles. Although the current 5G standard specifies the necessary methods, many deployments have not completely implemented them, and they may also be too sluggish to be used in some highly dynamic applications. It will be important to use non-Gaussian models and novel techniques to account for dependency in time and space [9].

11.2.5 NETWORK ENERGY EFFICIENCY

Energy expenditure is now one of society's top concerns. As a major consumer as well as a tool for lowering consumption, whether at home or by facilitating distant meetings, telecommunications networks have a special role in this context. Energy is the resource that occasionally determines the life of the network in applications like IoT, when batteries cannot be replenished. Energy conservation is always a key concern, which presents difficulties for network organization and hardware design. The development of fully autonomous nodes with gathered energy is a significant problem, as is computation for nearby sensors. The energy consumption of devices may be considerably reduced through new hardware design, for example by using artificial neural networks or spiking neural networks [10].

11.2.6 CO-EXISTENCE

Modern cellular networks are being created with a variety of traffic kinds and vastly varying service requirements in mind. Numerous licensed and unlicensed radio access methods will be used to guarantee connectivity. It is yet unclear how these technologies will cohabit, both for coexistence between various 5G services and between 5G and other applications, such as Wi-Fi. Inter-network interference, particularly in the industrial scientific and medical (ISM) bands, might become a serious issue [11]. Approaches based on duty-cycle or carrier detection are occasionally insufficient or unsuited to minimize the resulting interference. To design modified coexistence rules, better modeling and comprehension are required. It would also be beneficial to use more unlicensed bands, such as 60 GHz, in order to reduce the amount of traffic in the current bands. Further consideration of these options is necessary. Modern cellular networks are being created with a variety of traffic kinds and vastly varying service requirements in mind.

11.2.7 SECURITY AND PRIVACY

It is impossible to highlight significant issues that still require investigation without bringing up the privacy and security concerns. The integrity and verifiability of the received data must be safeguarded against an increasing number of security risks, including pure software assaults and access to the physical network structure. With so many inexpensive devices exchanging information, IoT also introduces new hazards. Due to the cheap cost, it is harder to safeguard this information, and there are more risks to users' privacy from both legitimately able-to-access organizations and data brokers as well as from criminals looking to violate privacy [12].

11.3 PREPARING FOR 6G

11.3.1 VISION

Despite the fact that 5G will offer higher quality of service (QoS) than 4G, it won't be able to create a fully autonomous, ubiquitous connectivity network that gives a fully immersive experience and offers everything as a service. However, it is

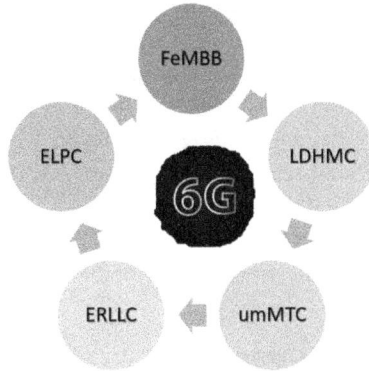

FIGURE 11.3 6G service classes.

anticipated that 6G will be able to simultaneously satisfy all of the demanding network requirements such as high reliability, enhanced capacity, improved efficiency, and low latency, given the comprehensively expected 2030 economic, technological, and environmental conditions. The Network 2030 focus group (ITU-T FGNET 2030) was established by the International Telecommunication Union (ITU) in July 2018 to examine how system technologies will advance for 2030 and beyond. Its 6G suggestions include new holographic media, services, network architecture, internet protocol (IP), and other concepts.

A brief description of some significant 6G wireless communication possibilities and applications are provided in the next section. Figure 11.3 highlights the different services classes of 6G.

11.3.2 DRIVING APPLICATIONS

11.3.2.1 Extended Reality (XR)/Virtual Reality (VR)

For 6G, extended reality (XR) will produce a number of game-changing applications throughout the augmented reality (AR) and VR spectrums. VR technology has been applied in several fields over the last few years, including the defense sector, healthcare, and education [13]. VR has the ability to fundamentally alter how we live and work since it offers wonderful immersive and interactive experiences while overcoming financial and safety barriers in the actual world. Because of its low cost, portability, and compact size, mobile VR is seen as the technology of the future. In a nutshell, 6G must be able to offer a combination of classic URLLC and eMBB with included perceptual aspects for XR services [14].

11.3.2.2 Industry Automation

The abbreviation "Industry 4.0" refers to the major transformation of the industrial sector due to the advent of new technologies such as robotics and application of AI. Industry 4.0 seeks to bring about change as opposed to the third phase of the industrial revolution, with automated manufacturing using electronics and information technology (IT) in industrial production with significantly improved efficiency based on a cyber-physical system. However, this vertical industry's uniqueness creates a number of difficulties that must also be taken into account, including the many use cases with stringent network performance requirements, the difficult propagation environment with possibly high interference that may necessitate seamless integration, the unique safety and security considerations, and the difficulties of using terminology from different industries [15].

Automation of factories, processes, human-machine interfaces, IT-based production, logistics and warehousing, monitoring, and preventive maintenance are some application areas.

11.3.2.3   Cellular Vehicle-to-Everything (C-V2X)

Autonomous cars have considerable potential for enhancing both traffic efficiency and road safety, and vehicle-to-everything (V2X) is a vital enabler. To offer communication capabilities with low latency, high reliability, and high throughput for vehicle-to-vehicle (V2V), vehicle-to-pedestrian (V2P), vehicle-to-infrastructure (V2I), and vehicle-to-cloud (V2C), a standardized V2X solution called cellular vehicle-to-everything (C-V2X) is recommended by the 3rd generation partnership project (3GPP). Vehicle intelligence and autonomous functionality will undoubtedly advance as AI spreads more quickly. The next-generation C-V2X would provide a systemic solution to support collaboration among the major players in an intelligent transportation system (ITS), in addition to expanding its fundamental capability from information delivery to information processing [16].

11.3.3   Wireless Brain-Computer Interactions (BCI)

Applications of brain-computer interactions (BCI) are frequently only used in hospitals where patients can use brain implants to control artificial limbs or other nearby computing devices. This technology will undergo a revolution thanks to recently developed wireless brain-computer implants and interfaces, which will also provide new use-case scenarios that call for a 6G connection. From permitting brain-controlled movie input to fully realized brain-controlled environments, these solutions offer a variety of services. Individuals will engage with their surroundings and other people utilizing discrete gadgets, some of which are worn on the body, some of which are implanted, and some of which are integrated into the environment around them, rather than on cellphones. As a result, people will be able to interact with their surroundings through gestures and send haptic messages to their loved ones [17].

11.3.4   Connected Robotics and Autonomous Systems (CRAS)

The eagerly anticipated deployment of connected robotics and autonomous systems (CRAS), which includes drone delivery systems, autonomous cars, autonomous drone swarms, vehicle platoons, and autonomous robotics, is a key driver for 6G systems. There is more going on with the launch of CRAS over the cellular domain than just "yet another short packet uplink IoE service." Instead, CRAS requires strict latency restrictions that are set by the control system, which may even require sending high definition (HD) maps via enhanced mobile broadband scenario of 5G [18, 19].

11.4   ENABLING TECHNOLOGIES

In 6G wireless communication networks, several unique enabling technologies will be employed in order to attain the aforementioned system performance criteria. A few of them are discussed in brief here.

11.4.1   NEW WAVEFORMS

The graph of a signal shape as a function of time in the physical media is called a waveform. A flexible waveform must take into account a number of factors, including peak to average power ratio (PAPR), spectral efficiency (SE), resistance to time/frequency dispersions, and time/frequency localization. Even though it has drawbacks, including a high PAPR, orthogonal-frequency division multiplexing (OFDM) has emerged as the dominant waveform for mobile communication since 4G. In order to lessen out-of-band radiation, several types of OFDM waveform schemes are being researched for the 5G standard. When selecting a new waveform, it is important to consider the design of the frame structure, flexibility of the parameter selection, complexity of the signal processing technique, and other elements [5, 20].

11.4.2   MULTIPLE ACCESS

Multiple access approaches have been acknowledged as significant turning points in the migration of cellular networks. Orthogonal multiple access (OMA) has been used in previous generations of mobile networks. In the time, frequency, or code domains (such as time slots, sub-carriers, and spreading codes), orthogonal bandwidth resource blocks are created and then orthogonally allotted to users, with each resource block being occupied by a single user. OMA's success in previous-generation mobile networks is largely attributable to the ease with which it may be deployed, even though it is well known that OMA's spectrum efficiency is less than ideal. Known as NOMA (non-orthogonal multiple access), it is a significant shift for the creation of numerous access methods for the future. The main goal of NOMA is to promote spectrum sharing among mobile users. It outperforms OMA in terms of spectral efficiency gain by making use of users' varied QoS needs or dynamic channel circumstances [21].

11.4.3   ADVANCED CHANNEL CODING AND MODULATION

A radio link can operate effectively near to its channel capacity by using channel coding and modulation, two fundamental physical layer technologies, which also make the signal waveforms compatible with RF (radio frequency) and baseband components at transmitters and receivers. In a wide sense, channel coding and modulation may be related to waveform or even multiple access. The three operational modes specified by the 5G standards are the eMBB mode, the uRLLC mode designed for various mission-critical applications, and lastly the mMTC for industrial and IoT applications. Naturally, the requirements of these three modes vary somewhat, particularly in terms of the acceptable latency. The performance parameters for advance channel codes are the error-correction capabilities, flexible code-rate, and the ability to support hybrid automatic repeat request (HARQ), low-density parity-check code (LDPC) are selected for enhanced mobile broadband (eMBB) data channel of 5G NR, because they are suitable for HARQ while being extremely adaptable in terms of their code rates, code lengths, and decoding latency. Polar codes, in contrast, have gained popularity for the low-delay control channels, and are adopted as the 5G control channel scheme [22].

11.4.4 AI FOR SPECTRUM SHARING

In-depth information exchange across the systems is often necessary for the creation of spectrum sharing solutions. Dynamic spectrum management will be challenging to accomplish, as 6G network settings get more complicated. To address these issues, dynamic spectrum management is increasingly supported by AI, which is a powerful enabler to tackle these challenges. For instance, when operating in the unlicensed spectrum, the LTE network can use deep reinforcement learning (DRL) to learn the Wi-Fi traffic and create an appropriate protocol to support equitable coexistence.

11.4.5 VISIBLE LIGHT COMMUNICATIONS

The most promising next-generation lighting technology is anticipated to be the laser diode (LD) phosphor conversion lighting technology, which exceeds conventional LEDs in terms of brightness, efficiency, and illuminating range. It is well known that the LD may be adjusted very quickly and has a wider modulated bandwidth than LED. With a proven speed of 28.8 Gbps and a theoretical speed of 100 Gbps, LD-based visible light communications (VLC) systems are better suited for ultra-high data density (uHDD) services in 6G networks. So, for 6G, an enhanced VLC (eVLC) based on LD lighting technology is suggested. On the other hand, compared to an LED with a comparable lumen output, the LD's beam angle will be between one-fifth and one-tenth of that. Due to the narrow patch of light, the LD-based illuminator generates output with sharp edges that is around ten times sharper than LED illumination. Additionally, larger distances will be supported by the LD lighting's extremely high directivity [23].

11.4.6 ENERGY EFFICIENT AND LOW POWER NETWORK

The world is heading towards analytic solutions for every sphere of technology that heavily rely on data correlation. It is now true that 5G networks use more power than 4G networks while providing a better bandwidth. Energy-efficient computing in the 6G network era entails greater resource utilization and reduced energy usage. Building on software upgrades for network operations is the logical strategy for preparing 6G networks for these uncharted new traffic requirements. Building domain-specific computer engines for 6G, which would drastically cut energy usage, is therefore the only answer currently foreseen. In a virtualized radio access network, this will enable unfettered adaptation to a relatively new and emerging necessity of domain-specific chips. Also, the edge computing paradigm is thought to be a fresh approach to computing that may successfully spread the strain on cloud computing. [10, 24].

11.5 FUTURE DIRECTION

In comparison to 5G, the technologies under discussion would significantly alter the physical layer of mobile communication networks. Numerous topics are now the subject of scientific inquiry. Given the rapid advancement of cloud computing, network function virtualization, and the concept of software-defined networks (SDNs) under the 5G umbrella, a paradigm shift towards 6G mobile communications is visibly clear. Furthermore, they do actually reflect how far a nation has come in key fields of science and technology. The cutting-edge technologies must yet undergo further research and development before they can be used for engineering. Although much has been achieved over the past five years, much more has to be done to allow networks to provide a totally new encounter with human interaction or human interaction with the physical environment. Wireless is far from mature since its requirements and applications are always changing and becoming more difficult. The twin-component ultra-reliable low-latency mode of the 5G system illustrates the evolutionary paradigm change away from pure single-component architectures focused on bandwidth efficiency, power efficiency, or delay optimization. The five primary areas of research in this field at early stage of their evolution are as follows.

11.5.1 Advanced Wireless Technology

With the integration of satellite, underwater, and aerial communication networks, it is projected that the next 6G wireless network would go above the existing terrestrial cellular network to develop a vertical 3D network, also known as 3DNet or SkyNet. As a result, one of the main research topics is how to manage these new heterogeneous vertical/horizontal enormous ultra-dense networks (UDNs).

11.5.2 Software-defined Networks (SDN)/Network
Function Virtualization (NFV)

Software-defined networks (SDN) and network function virtualization (NFV) are described as the essential foundations that enable the wide variety of key performance indicators (KPIs) for the new 5G use cases in a cost-effective manner by the 5G Infrastructure Public Private Partnership (5GPP20). SDN research focuses on both autonomous network administration and the optimization of traffic management to meet the demanding requirements of next-generation networks. SDN offers the network's flexibility to change via a centralized network control system, a network architecture that does away with the vertical integration seen in conventional networks. Load balancing, multi-path routing, dynamic real-time automated network management, quality management of service routes, and automatic repair of faulty routes are all included in this network [25].

11.5.3 MM-WAVE TRANSMISSION

A benefit of an mm-wave system operating generally at above 20 GHz is that it can provide Gbps-level transmission. This is accomplished, nevertheless, at the cost of significant path loss and sensitivity to obstacles. Beamforming based on high-gain beams may be used as a viable treatment to reduce the path loss [20].

11.5.4 TERA HERTZ (THZ) TRANSMISSION

The carrier frequencies of the THz band have even larger bandwidth than the mm-wave spectrum. They range from 0.3–3THz; as a result, they will be able to accommodate terabits per second (Tbps) transmission rates. However, because of intense molecular absorption, this spectrum has a substantially larger path loss than mm-wave carriers. Accurate channel models for THz communications must be established in order to develop effective wireless communication systems in the THz range. THz channel modelling must take into account a number of distinctive characteristics that are not present in lower-frequency channels, such as the mutual coupling effect and the very frequency-selective path loss [26].

11.6   NEW MULTIUSER TRANSMISSION SCHEMES

11.6.1   3D BEAMFORMING

This idea comes from sectorized antennas that employ advanced beamforming methods and antenna layouts to serve customers who are travelling at various angles. The use of high-gain beamforming is essential because mm-wave carriers have severe path loss. Even more essential is high-gain beamforming for THz carriers that use pencil beams [26].

11.6.2   NON-ORTHOGONAL MULTIPLE ACCESS (NOMA)

In contrast to OMA, NOMA makes use of a resource block to serve numerous users at once by using the power domain dimension. In some resource domains, the transmitter will multiplex a number of desirable signals into a single resource slot, while the receiver uses successive interference cancellation (SIC) to reduce the interference brought on by signals from other users. In the most popular power-domain NOMA, two or more users' signals are superimposed at a particular power, and SIC is used to identify the strongest signal while classifying the weaker signals as interference. The weaker signal is then left behind after the remodulated signal has been subtracted from the aggregate signal. Code-domain NOMA also employs similar SIC-assisted techniques [13, 27].

11.6.3   Full Duplex (FD)

Using this technique, uplink and downlink transmission can happen at the same frequency and at the same time window. However, a fundamental problem that necessitates sophisticated interference avoidance approaches is the interference between the high broadcast power and the low received power. There is a chance for full duplex (FD) solutions to boost spectrum efficiency by using mm-wave/THz MU-MIMO beamforming in both terrestrial and aerial networks with the advent of the new 6G architectures and technologies.

11.6.4   Coordinated Multi-point (CoMP)

A number of base stations can transmit at once to a single receiver using this approach. Additionally, as an advantage of mm-wave/THz beamforming, ultra-massive MIMO (UM-MIMO) CoMP has the capacity to increase network throughput. There are major synchronization issues that prevent a wide-scale CoMP rollout since the receiver can only be fully synced with a single base station [27].

11.6.5   Reconfigurable Intelligent Surfaces (RISs)

Reconfigurable intelligent surfaces (RISs) are potential methods for increasing data speeds, expanding the service area, and consuming less energy. Numerous metamaterial components make up the RIS, which can reflect incoming waves and change their phases without the need for sophisticated signal processing. It will work well to use RIS, which will be mounted on building doors and windows to reflect received signal without interference [28].

Figure 11.4, demonstrates the multi-user, co-ordinated network of 6G.

11.6.6   Next-Generation Packet Core Networks

The serving gateways (SGWs), home subscriber server (HSS) units, mobility "management entity (MME), and other specialized hardware and software make up the traditional mobile core network. Although the typical core network has distinct control and data connection levels, the switch and router should nonetheless handle packets between these two tiers simultaneously. Given the rapid development of SDN and NFV, the mobile core network is the natural foundation for accommodating the control and data link layers, which give rise to the possible study issue of the so-called virtualized evolving packet core.

11.6.7   Next-Generation Mobile Network Architecture

SDN and NFV have made it possible for next-generation mobile networks to be designed and managed with a high degree of flexibility, intelligence, and automation, including cutting-edge mobile cloud and edge computing. A significant difficulty is adaptively assigning computer resources to the cloud and edge. Fog computing and mobile edge computing (MEC) are both significant next-generation

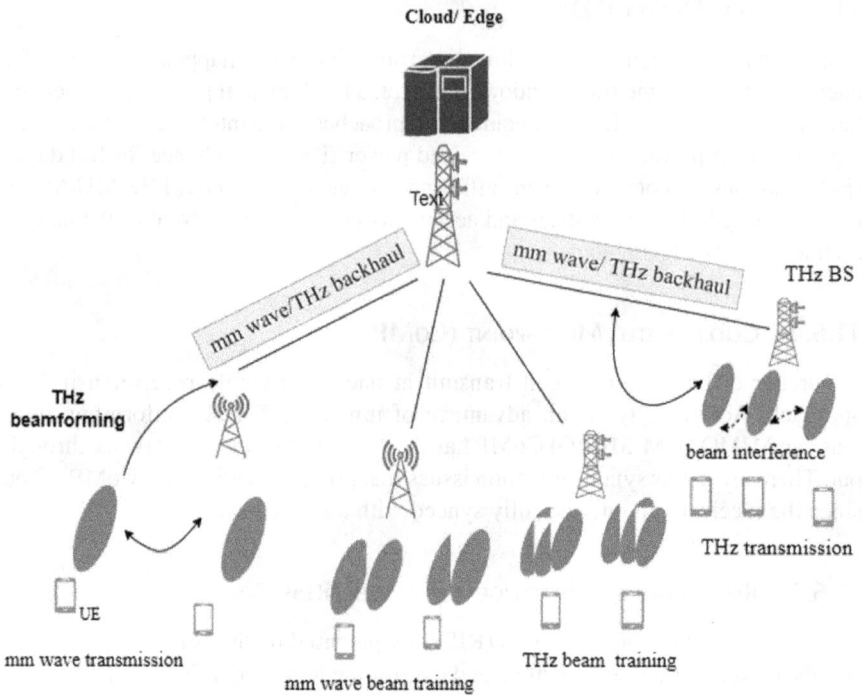

FIGURE 11.4   Network architecture of 6G.

network designs that have the ability to reduce service latency and increase spectral efficiency and QoS.

11.6.8   IoT Access and Sensing

IoT is a network made up of physical things or their associated hardware and software that gather data from various networks made up of heterogeneous sensor devices and controllers. Just a few instances of how various IoT networks necessitate diverse kinds of sensors and transmission states, are the Internet of Everything (IoE), the network of personal wearables, industrial IoT (IIoT), intelligent home services, and even the underwater internet. Modern IoT applications that demand mobility, such as smart cities, health care, smart cars, hospitals, and post-emergency networks, have already adopted a broad variety of data collecting devices [29].

11.6.9   Vehicular Networks

The growing need for high-performance Internet of Vehicles (IoV) services is a distinctive IoT application case that is fueling major research. One of the largest challenges comes from connected autonomous vehicles (CAVs), which rely on integrated sensing, control, and communications. The IoT devices and sensors are carried by the vehicles in order to recognize and track objects like people, other automobiles, and traffic signals. Additionally, monitoring the flow of traffic on the roads is crucial. V2X also facilitates wireless data transfer between moving objects and other linked devices [30].

11.7   WIRELESS SENSING AND POSITIONING

11.7.1   OUTDOOR ENVIRONMENT POSITIONING

The global positioning system (GPS) is the most used outdoor positioning system; however, due to its significant signal loss, it has a low location accuracy and a small area of coverage. Its flaws might be improved by using the signals obtained from cellular base stations, such as for the position monitoring of automobiles and unmanned aerial vehicles (UAVs). We are able to create appealing outdoor apps that revolve on local sites of interest by relying on advanced AI algorithms.

11.7.2   INDOOR ENVIRONMENT POSITIONING

Due to the limited inside range of the existing GPS system, indoor location systems generally rely on Wi-Fi signal strength evaluations and "fingerprinting" of pre-recorded RF maps. However, the dependability of wireless signals and the time-consuming development of large databases have a significant impact on the accuracy of fingerprint-based indoor locating technologies. The use of spatial skeleton databases made from indoor map data may help this investigation [32].

11.7.3   INDOOR WIRELESS DETECTION

The application scenarios for smart homes, green buildings, factory production, and healthcare will all benefit greatly from the quick development of device-free indoor positioning. These advanced detection methods for pedestrian route tracking, presence recognition, motion detection, and vitality detection will also find novel applications [32–34].

11.7.4   DEEP LEARNING IN 6G NETWORKS

AI-enabled reinforcement learning, supervised learning, unsupervised learning, and deep learning have all become more popular techniques for resolving challenging networking and communications issues. The control of radio interference, resource assignment, optimization of numerous parameters, prediction of network traffic, allocation of computing resources, and flexible network function setup are more detailed examples. In supervised learning, a deep layered neural network (NN) is made up of fixed-size inputs and ground truth labels. Unsupervised learning, which makes use of the correlation between dataset samples, does not, however, need labeling. An agent will interact with the environment during reinforcement learning, and the model will be updated depending on the associated rewards.

11.8 CONCLUSION

The need for a 6G network is emphasized in this chapter since 5G is unable to meet the complex and demanding criteria of the next applications. The prospective 6G major technologies, which comprise both innovative and more advanced technology, are thus the main emphasis. In comparison to 5G, the disruptive technologies under discussion would significantly alter the physical layer of mobile communication networks. As we transition into the next generation of mobile networks, 5G represents a crucial inflection point, where sophisticated machine-to-machin" and machine-to-human interactions are expected to fundamentally alter how we live and work. Numerous elements are currently being researched scientifically. But they do in fact demonstrate a country's level of scientific and technological advancements. For the advanced technologies to be practical in engineering, further physical layer research and development must be done. In 10 years (2030), the 6G network will support more than 1,000 wireless nodes per person, deliver terabit rates per second, and offer immediate holographic connectivity wherever you are. In a millisecond-connected, completely data-driven world, people and things will be globally connected in the future.

REFERENCES

1. 3rd Generation Partnership Project (3GPP) TS 29.500, 5G System: Technical realization of service-based architecture, Stage 3, v1.0.0, Mar. 2018
2. Shafi, M., Molisch, A. F., Smith, P. J., Haustein, T., Zhu, P., De Silva, P., ... & Wunder, G. (2017). 5G: A tutorial overview of standards, trials, challenges, deployment, and practice. *IEEE Journal on Selected Areas in Communications*, 35(6), 1201–1221.
3. Yuan, Y., Zhao, Y., Zong, B., & Parolari, S. (2020). Potential key technologies for 6G mobile communications. *Science China Information Sciences*, 63, 1–19.
4. Saad, W., Bennis, M., & Chen, M. (2019). A vision of 6G wireless systems: Applications, trends, technologies, and open research problems. *IEEE Network, 34*(3), 134–142.
5. Banafaa, M., Shayea, I., Din, J., Azmi, M. H., Alashbi, A., Daradkeh, Y. I., & Alhammadi, A. (2022). 6G mobile communication technology: Requirements, targets, applications, challenges, advantages, and opportunities. *Alexandria Engineering Journal, 64*, 245–274.
6. Jabagi, N., Park, A., & Kietzmann, J. (2020). The 5G revolution: Expectations versus reality. *IT Professional, 22*(6), 8–15.
7. Indoonundon, M., & Pawan Fowdur, T. (2021). Overview of the challenges and solutions for 5G channel coding schemes. *Journal of Information and Telecommunication, 5*(4), 460–483.
8. Erunkulu, O. O., Zungeru, A. M., Lebekwe, C. K., Mosalaosi, M., & Chuma, J. M. (2021). 5G mobile communication applications: A survey and comparison of use cases. *IEEE Access, 9*, 97251–97295.
9. Fourati, H., Maaloul, R., & Chaari, L. (2021). A survey of 5G network systems: challenges and machine learning approaches. *International Journal of Machine Learning and Cybernetics, 12*, 385–431.
10. Usama, M., & Erol-Kantarci, M. (2019). A survey on recent trends and open issues in energy efficiency of 5G. *Sensors*, 19(14), 3126.
11. Burkacky, O., Hoffmann, A., Lingemann, S., & Simon, M. (2020). The 5G era: New horizons for advanced-electronics and industrial companies.

12. Bartolín-Arnau, L. M., Vera-Pérez, J., Sempere-Payá, V. M., & Silvestre-Blanes, J. (2023, April). Private 5G Networks for Cyber-Physical Control Applications in Vertical Domains. In 2023 IEEE 19th *International Conference on Factory Communication Systems (WFCS)* (pp. 1–4). IEEE.

13. Shen, L. H., Feng, K. T., & Hanzo, L. (2023). Five facets of 6G: Research challenges and opportunities. *ACM Computing Surveys, 55*(11), 1–39.

14. Salameh, A. I., & El Tarhuni, M. (2022). From 5G to 6G—challenges, technologies, and applications. *Future Internet, 14*(4), 117.

15. Zhao, Y., Zhao, J., Zhai, W., Sun, S., Niyato, D., & Lam, K. Y. (2021). A survey of 6G wireless communications: Emerging technologies. In *Advances in Information and Communication: Proceedings of the 2021 Future of Information and Communication Conference (FICC), Volume 1* (pp. 150–170). Springer International Publishing.

16. Chen, S., Hu, J., Shi, Y., Zhao, L., & Li, W. (2020). A vision of C-V2X: Technologies, field testing, and challenges with Chinese development. *IEEE Internet of Things Journal, 7*(5), 3872–3881.

17. Kamruzzaman, M. M. (2022). Key technologies, applications and trends of internet of things for energy-efficient 6G wireless communication in smart cities. *Energies, 15*(15), 5608.

18. Chen, S., Liang, Y. C., Sun, S., Kang, S., Cheng, W., & Peng, M. (2020). Vision, requirements, and technology trend of 6G: How to tackle the challenges of system coverage, capacity, user data-rate and movement speed. *IEEE Wireless Communications, 27*(2), 218–228.

19. Adem, N., Benfaid, A., Harib, R., & Alarabi, A. (2021). How crucial is it for 6G networks to be autonomous? arXiv preprint arXiv:2106.06949.

20. Moltchanov, D., Sopin, E., Begishev, V., Samuylov, A., Koucheryavy, Y., & Samouylov, K. (2022). A tutorial on mathematical modeling of 5G/6G millimeter wave and terahertz cellular systems. *IEEE Communications Surveys & Tutorials.*

21. Zhao, Y., Zhai, W., Zhao, J., Zhang, T., Sun, S., Niyato, D., & Lam, K. Y. (2020). A comprehensive survey of 6g wireless communications. arXiv preprint arXiv:2101.03889.

22. Patil, M. V., Pawar, S., & Saquib, Z. (2020, April). Coding Techniques for 5G Networks: A Review. In 2020 3rd *International Conference on Communication System, Computing and IT Applications (CSCITA)* (pp. 208–213). IEEE.

23. Giordani, M., Polese, M., Mezzavilla, M., Rangan, S., & Zorzi, M. (2020). Toward 6G networks: Use cases and technologies. IEEE Communications Magazine, *58*(3), 55–61.

24. Hu, N., Tian, Z., Du, X., & Guizani, M. (2021). An energy-efficient in-network computing paradigm for 6G. IEEE Transactions on Green Communications and Networking, *5*(4), 1722–1733.

25. Hassebo, A. (2022, January). The road to 6G, Vision, Drivers, Trends, and Challenges. In 2022 IEEE 12th *Annual Computing and Communication Workshop and Conference (CCWC)* (pp. 1112–1116). IEEE.

26. Sopin, E., Moltchanov, D., Daraseliya, A., Koucheryavy, Y., & Gaidamaka, Y. (2022). User association and multi-connectivity strategies in joint terahertz and millimeter wave 6G systems. IEEE Transactions on Vehicular Technology, *71*(12), 12765–12781.

27. Saad, W. (2021). 6G wireless systems: Challenges and opportunities. *5G and Beyond: Fundamentals and Standards*, 201–229. https://doi.org/10.1007/978-3-030-58197-8_7

28. Liang, Y. C., Chen, J., Long, R., He, Z. Q., Lin, X., Huang, C., … & Di Renzo, M. (2021). Reconfigurable intelligent surfaces for smart wireless environments: channel estimation, system design and applications in 6G networks. *Science China Information Sciences, 64*, 1–21.

29. Li, J., Dang, S., Wen, M., Li, Q., Chen, Y., Huang, Y., & Shang, W. (2023). Index modulation multiple access for 6G communications: principles, applications, and challenges. *IEEE Network, 37*(1), 52–60.

30. Garcia, M. H. C., Molina-Galan, A., Boban, M., Gozalvez, J., Coll-Perales, B., Şahin, T., & Kousaridas, A. (2021). A tutorial on 5G NR V2X communications. *IEEE Communications Surveys & Tutorials, 23*(3), 1972–2026.

31. Imoize, A. L., Adedeji, O., Tandiya, N., & Shetty, S. (2021). 6G enabled smart infrastructure for sustainable society: Opportunities, challenges, and research roadmap. *Sensors, 21*(5), 1709

32. Ali, S., Saad, W., Rajatheva, N., Chang, K., Steinbach, D., Sliwa, B., ... & Malik, H. (2020). 6G white paper on machine learning in wireless communication networks. arXiv preprint arXiv:2004.13875.

33. Sangeetha, S. K. B., & Dhaya, R. (2022). Deep learning era for future 6G wireless communications—Theory, applications, and challenges. In R. Kanthavel, K. Ananthajothi, S. Balamurugan, & R. K. Ganesh (eds.), *Artificial intelligent techniques for wireless communication and networking* (pp. 105–119). https://doi.org/10.1002/9781119821809 .ch8

34. Gündüz, D., de Kerret, P., Sidiropoulos, N. D., Gesbert, D., Murthy, C. R., & van der Schaar, M. (2019). Machine learning in the air. *IEEE Journal on Selected Areas in Communications, 37*(10), 2184–2199.

12 Task Prediction and Optimal Offloading Decisions for Fog Computing Cloud Network

Dheeraj Kumar Sharma and Niraj Pratap Singh

12.1 INTRODUCTION

With the increasingly widespread use of digital devices such as smart mobiles, wearable technology, and automobiles that are empowered with different sensors, internet of things (IoT) networks, and mobile cloud computing networks, in which information is shared between linked products and systems, including such home automation, sustainable grid and smart manufacturing have gotten a lot of attention [1, 2]. Because of its flexible and efficient processing resources, cloud computing has also been used to process tasks from devices, notwithstanding the complexity and diversity of services [3]. On the other hand, the cloud servers at a distant location usually result in excessive latency and demand a lot of network capacity for the network infrastructure. In the past few years, the explosive increase of computation-intensive and delay-sensitive applications, such as multiplayer gaming, image or signal sensing, wearables, and real-time translators, has positioned mass data workload on mobile edge computing resources (MECs). Because of the constraints of mobile devices of processing, battery capacity, and storage, there is a growing trend to offload or move computer-intensive tasks to area networking computing systems. This method is known as computation offloading [4, 5]. It saves energy by reducing the energy consumed for storage and processing, resulting in longer battery life.

The cloud-based network topology may offer powerful processing capabilities without requiring terminal nodes to do their duties locally. Nonetheless, as the number of terminal nodes grows exponentially, including smartphones and tablets, autonomous cars, smart home appliances, and so on, vast amounts of data must be handled quickly [6–8]. This places a significant load on the cloud server's transmission connection, and the vast distance between the current node and the mobile cloud server causes unsatisfactory task processing delays, inadequate mobility support, and security concerns. As shown in Figure 12.1, a large number of fog nodes are projected to be distributed throughout the system in the future 5G system. Computation relayed networking and control with other services may be freely installed on such

DOI: 10.1201/9781003407836-12

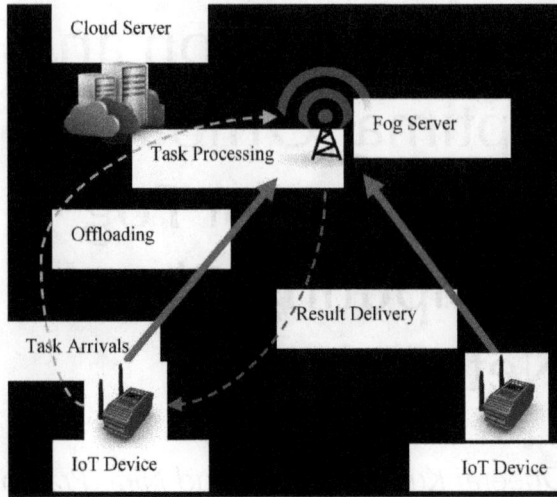

FIGURE 12.1 Fog-enabled task offloading network architecture.

ubiquitous nodes, resulting in a user-centric network that can ultimately serve all types of cloud server requirements. As depicted in Figure 12.1, the fog nodes are used to process the offloaded tasks from IoT devices. This fog computing (FC)-based task offloading system might have superior delayed performance [9] by using numerous neighboring fog nodes surrounding the terminal node. Figure 12.1 depicts the architecture of fog-enabled task offloading network.

In an FC network, however, the portion of the fog nodes might be dedicated localized servers with powerful processing capabilities and high power consumption [10]. Task offloading schemes might prioritize task delays, or overall energy usage might place undue strain on the fog nodes closest to the terminal node. Given the diverse computing capabilities and long-term viability of fog nodes, achieving fair energy usage among fog nodes is critical for achieving excellent network performance [11]. The choice to offload a wireless device's computational tasks to MEC servers is considered very carefully. If computing tasks are vigorously offloaded to the fog-cloud server, the uplink communication networks will become severely congested, resulting in a significant delay in completing computer tasks [12]. As a result, in order to maximize the benefits of computation offloading, this study maintains a close check on the fairness of the nodes in the network to whom the computational tasks are offloaded. Each fog node's performance indices, which include task delays and energy usage with response time, are calculated [13].

The main contribution of the chapter includes

a. The presentation of an optimization-based algorithm, termed as long short-term memory—grey wolf optimization (LSTM-GWO), for task predicting and optimum offloading.
b. The first known implementation of an LSTM-GWO algorithm for task prediction and optimum offloading as per the literature survey.

   c. In comparison to the ACO and PSO algorithms, the suggested LSTM-GWO offloading method maintains reduced average response time. LSTM-GWO average response time ranged from 3–6 seconds but ACO and PSO's reaction times increased from 79–100 seconds as the number of nodes increased from 0–2000.

   d. The suggested LSTM-GWO method is faster than the PSO and ACO algorithms in terms of convergence. The optimal value is attained after iteration number is 60.

   e. The suggested LSTM-GWO task offloading algorithm obtains increased load imbalance standard deviation as compared to ACO and PSO algorithms.

   f. In the proposed method, as the number of fog nodes increases while maintaining the same parameter settings, the suggested LSTM-GWO algorithm retains fewer values in account of degree of inequality (DI), indicating that the proposed technique efficiently balances the workload across the fog nodes.

12.2 LITERATURE REVIEW

This section includes the related works of different algorithms used for task offloading and optimal offloading decision.

X. Chen et al. [14] proposed a combination strategy to enhance allocation of resources and task-offloading for augmented reality (AR) technologies is presented in either a single or multi-MEC system to minimize energy utilization. It contains a quantitative simulating model for the convergence features, incentives, and energy usage, illustrating how the proposed solution surpasses the competition in terms of customer energy usage. The typical user energy usage of the multi-MEC method is 71.0 percent lower than those of the single-based MEC algorithm.

Singh et al. [15] studied the challenges of energy-effective task offloading for IoT systems. The authors proposed an efficient task allocating across localized connected devices, edge servers, and the cloud in task to reduce energy usage and end-to-end latencies. FC is quickly becoming a popular option. It consists of moving data processing tasks to the network's edge. FC, on the other hand, has its own set of obstacles that must be solved in order to create efficient and successful solutions. The energy-efficient task offloading method (EETOS) is then used to address the issue, which is based on the Levy-flight moth flame optimizing (LMFO) strategy. In comparison to previous algorithms, EETOS decreases energy usage.

For small-cell networks, L. Yang et al. [16] provided a distributive computational offloading approach that is combined with MEC. The potential game-based offloading algorithm (PGOA) was created as a consequence in order to acquire a Nash Equilibrium (NE) with restricted updates. In addition, they evaluated the PGOA's shortest iteration durations to get the worst-case performance and the efficiency ratio measurements to quantify the offered approach's computational efficiency. Furthermore, simulation findings show that the PGOA could not only minimize offloading latency and also optimize overhead effectiveness while enabling each mobile device (MD) to pursue its own objectives when contrasted to specific other contemporary computing approaches. The findings show that the PGOA approach can save up to 11 percent of the time.

Muhammad et al. [17] suggested middleware in FC because it delivers edge and FC services, as well as the ability to entertain delay-sensitive activities successfully. The issue occurs when a choice must be made about what should be offloaded, computed, or implemented and how to perform offloading. The suggested logistical regression framework achieves 86.01 percent accuracy when contrasted with different techniques, demonstrating the validity of the proposed model and trust in the predicted task-offloading strategy by ensuring continuous manner and dependability.

Z. Cao et al. [18] investigated the problem of multichannel availability and task offloading in the MEC-enabled Industrial-Revolution-4.0 and explained it from the perspective of a multi-agent system. To address this issue, a novel deep-reinforcement technique approach was suggested. Edge devices (EDs) may communicate with one another, which reduces computation time and improves channel allocating success rates. Simulation studies reveal that the proposed technique may reduce calculation latencies by 33.30 percent.

Nguyen et al. [19] proposed reinforcement learning (RL)-based algorithms, which enable mobile users (Mus) to find the best offloading options. Experimentation and modeling data showed that the suggested RL-based offload techniques considerably increase safety and confidentiality, enhance sustainability, and decrease computation latency with negligible offload costs compared to the benchmark offload systems. The quantitative findings show that the DRL-based task offloading (DRLO) approach outperforms existing baseline methods in terms of improving offload cost efficiency in multi-user scenarios. Offloading costs for the DRLOs are lower than other techniques. The efficiency of the DRLO-based information task offload approach is shown by these results.

Y. Kyung et al. [20] proposed an analytical framework for the opportunistic type of offloading likelihood, in which the task may be offloaded to the opportunistic fog nodes (OFN). The analytic model is validated, and the effectiveness of the opportunistic offload probabilities is shown using extensive simulation data. Offloading of computing workloads to the fog node, which is co-located with the base station (BS), is possible in the FC architecture. However, since IoT systems inside the BS's coverage area might offload a large number of activities at a time, the FN can become used, causing scalability concerns owing to limited processing capabilities.

Sun, Ming et al. [21] proposed application-driven task offloading strategy (ATOS), a novel deep reinforcement learning (DRL)-based task offloading technique that involves an initial screening phase. Examine the features of DRL-based tasks and suggest the preliminary sorting mechanism (PSM) heuristic approach for finding the processing order of concurrent sub-tasks. Several experiments were also undertaken to test ATOS's convergence and efficiency. When compared to conventional methods, ATOS might just save up to 64.50 percent on total overhead costs.

The recent advances in task offloading are presented in Table 12.1.

TABLE 12.1
Recent Advances in Task Offloading

References	Technique	Result
[14]	DL based	Energy consumption is 71% lower.
		Energy consumption varied from 0.024–0.02.
[15]	Levy-flight moth flame optimization (LMFO)	Lower energy usage by 22.0, 25.0, and 29.0%.
[16]	Distributed computation offloading strategy	11% reduction in saving delays.
[17]	Logistics Regression based offloading scheme	Logistic regression model acquiring 86% accuracy.
[18]	DL based task offloading	Improvement in latency by 33.38%.
[19]	Reinforcement learning (RL)	Offloading costs for the DRLOs decreased 18.70%, 57.0%, and 64.0%.
[20]	Hybrid offloading (HO)	HO accounts for 73.0% of energy optimization.
[21]	Application-driven task offloading	Optimization rates are increasing by 78.30% on average

12.3 METHODOLOGY

In this section, discussion on system model of FC and local computing is done. It also contains problem formulation, task prediction model, and optimal offloading decision.

12.3.1 SYSTEM MODEL

In the present work, we assume a multi-edge or multi-fog system with one cloud server, one wireless access point (AP), and N wireless devices, represented with the set N = {1, 2..., n} in Figure 12.1. A fiber optics cable connects the access point and the fog server, and the transmission delay/response time may be disregarded. A number of tasks are allotted to each wireless device. It is supposed that every wireless device has M independent tasks, signified with a set T = {1, 2..., t}, without losing generality. These tasks should be completed locally or forwarded to fog server via AP. The workload of user n's t^{th} task is signified by $D_{n,t}$. Every task 't' of 'n' is decided using deep learning approach to whether it is offloaded or executed locally, and this condition is represented as binary decision $D_{n,t} \in \{0, 1\}$. If it is offloaded, then $D_{n,t}$ is 1 otherwise, 0. The thorough operation shows the fundamental processes of FC as well as local computing given below.

12.3.2 FOG COMPUTING

When a task is offloaded to the cloud server, $D_{n,t}$ is sent to the AP over wireless links, where it is transmitted and analyzed. Since the information amount of feedback is minimal in general, we ignore the energy usage and latency during fog server and communicates computed results back to a wireless device. The energy used by wireless devices for transferring their workload to the fog server is denoted by $E_{n,t}$, while the cost of energy for information processing at the fog server is modeled as a linear combination of workload $D_{n,t}$, and it is represented mathematically as

$$E_{nt}^{total} = E_{nt} + \alpha D_{nt}, \tag{12.1}$$

where, α = weight of energy utilization.

The total energy is evaluated by adding the energy cost for sending tasks by wireless devices and the processing cost at the fog server. While offloading task, there exists some delay/response time, and it is mathematically evaluated as

$$TD_{nt}^{total} = \frac{D_{nt}}{b_n}, \tag{12.2}$$

where, b_n = bandwidth allotted to user 'n'.

Further, processing delay/response time at the fog server is calculated mathematically as

$$TD_{nt}^{fog} = \frac{D_{nt}}{f_r}, \tag{12.3}$$

where, f_r = processing rate at fog server.

Then, the overall delay/response time is calculated as

$$TD_n^{total} = \sum_{t=1}^{T} TD_{nt}^{total} + TD_{nt}^{fog}. \tag{12.4}$$

We consider that fog server could only initiate processing user n's tth task after fog received task completely and the fog server might only start redirect data output when full task t is finished.

12.3.3 LOCAL COMPUTING

Furthermore, we simulate situation such that user selects to complete task locally. We utilize E_n^l to represent the user n's local energy usage per data bit. As a result, the energy usage of user n for doing task 't' locally is represented by

$$E_{nt}^{local} = D_{nt} E_n^l. \tag{12.5}$$

Meanwhile, the localized processed time/data bit for user n is denoted as T^{local}. As a consequence, the total processing time required for user n to complete task t is

$$T_{nt}^{local} = D_{nt} T^{local}. \tag{12.6}$$

As a result of user n's offloading choice D_{nt}, total time it takes for users n to complete their activities locally is provided by

$$T_n^{local} = \sum_{t=1}^{T} T_{nt}^{local}(1 - o_{nt}).$$ (12.7)

12.3.4 PROBLEM FORMULATION

The main issue in task offloading is that total delay/response time can be represented as a system characteristic as $S_c\{d, o, b\}$, which is collectively weighted as the energy consumption and delay/response time characteristics. Mathematically, this constraint is represented as

$$S\{d,o,b\} = \sum_{n=1}^{N}(T_n^{local} + E_{nt}^{local} * o_{nt}) + \beta * \max(T_{nt}^{local}, T_n^{local}),$$ (12.8)

where, d = $\{D_{nt} \mid n \in N, t \in T\}$, o = $\{o_{nt} \mid n \in N, t \in T\}$, b = $\{b_n \mid n \in N\}$ and β = weight of energy consumption and task completion. Then, multi-objective optimization is used to minimize the system characteristic as $S_c\{d, o, b\}$ that is formulated to each user n's offloading decisions, o_{nt} and the bandwidth allocations b_n. For user n's task transmission, that is mathematically represented as

$$fun\{x\} = minimize\ (S\{d,o,b\})$$

$$Subject\ to\ \sum_{n=1}^{N} b_n \leq B$$

$$b_n \geq 0, \forall n \in N$$ (12.9)

$$o_{nt} \in \{0,1\}$$

The constraint tells that overall uplink bandwidth allotment for every user is less than C. b_n can be 0 or 1. For o_{nt}, 0 or 1 is provided.

12.4 TASK PREDICTION MODEL AND OPTIMAL OFFLOADING DECISION

12.4.1 TASK PREDICTION MODEL

After task creation, a decision procedure is necessary for computationally offloading technologies, and there'll be a time delay/response time between task generation and offloading choice. Even though task creation is a dynamically made and unpredictable process, given the work's long-term characteristics, this will have a significant relationship with time. As a result, we could forecast the tasks which will be created within the next networking timestamp of devices depending on their history, as well as load of the service packages ahead of time before the actual tasks arrive (e.g., allocating the most appropriate computational nodes prior with the decision model). For example, insert D_{nt} as the input data into the trained long short-term memory (LSTM) estimation method to forecast a subsequent time slot task information t for every node n with task data D_{nt}. As a result, the predictor model's optimizing target is *minimize* $S\{d,o,b\}$. In an actual case, as illustrated in Figure 12.2, we may use the prediction system to anticipate the data for a future task, make a choice, and allocate computing resources again for the task. If the difference between both the actual task and the scheduled task is within the acceptable threshold, whenever the real task comes, the task is immediately offloaded and calculated using the allocated decision data. Alternatively, the decision process is used to make the offloading choice, and the novel task's data is contributed to the historical information as a training set. The weights and biases for every gate in the LSTM network are changed during training to increase the prediction model's accuracy. Figure 12.2 shows the task prediction flowchart.

Whenever the server's load is not addressed in a fog-cloud computing system, incorrect offloading choices might result in load imbalance. The server's monitor keeps track of system real-time performance parameters. Considering input data, we use the optimization model to balance the server load distribution.

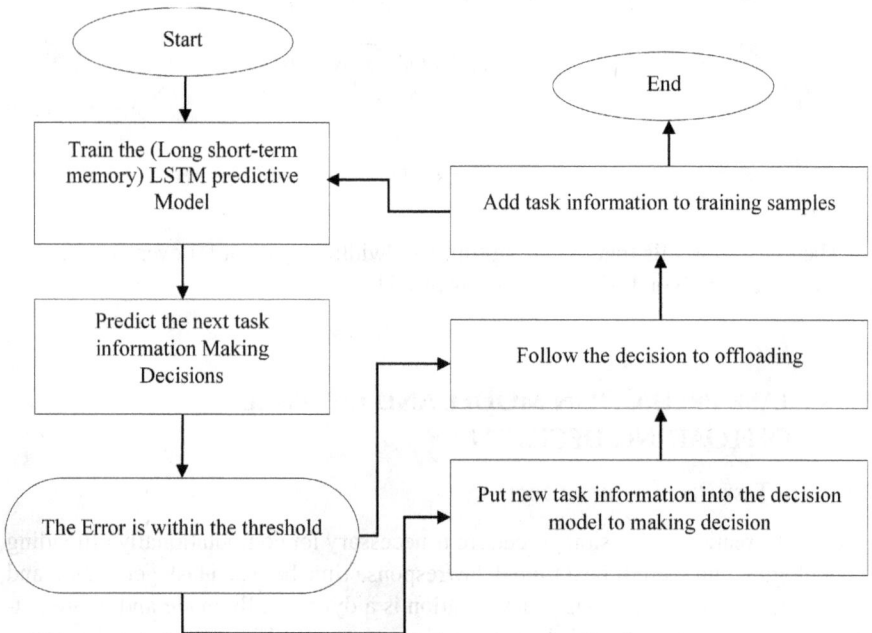

FIGURE 12.2 Flowchart of task prediction.

12.4.2 OPTIMAL OFFLOADING DECISION

The grey wolf optimization algorithm optimizes the grey wolf population by computationally replicating the monitoring, encircling, chasing, and attacking processes. The three phases of the GW hunting process include hierarchical social structuring, surrounding the prey, and attacking the prey.

Social structure grey wolves are top-of-the-food-chain social canids with a closed system of social authority. The best solution is denoted by the α; the second-best solution is represented by the β; the third-best solution is indicated by the δ, and the other solutions are denoted by the ω. Its social hierarchy is dominating.

Surrounding the prey during the hunting, grey wolves surround their prey; the following formula is used to represent encircling behavior quantitatively.

$$X(t+1) = X_p(t) - A \cdot \left| C \cdot X_p(t) - X(t) \right|$$
$$A = 2a \cdot r_1 - a$$
$$C = 2 \cdot r_2 \tag{12.10}$$
$$a = 2 - 2\frac{t}{\text{max\_iter}}$$

where X is the grey wolf's position vector, Xp is the prey's positioning vector, t is the current iteration, A and C are coefficient vectors, r1 and r2 are random vectors in $[0, 1]^{\wedge n}$, and is the distance control parameter, and its value reduces linearly from 2 to 0 over iterations, and Max_{iter} is the highest iterations.

Attacking the prey: Grey wolves have the capacity to determine the locations of probable prey, and the hunt is guided mainly by α, β, δ wolves. The best three wolves (α, β, δ) in the present group are kept within every iteration, while the locations of other search agents are modified based on their position data.

In this context, the simultaneous equations are presented.

$$X_1 = X_\alpha - A_1 \left| C_1 \cdot X_\alpha - X \right|$$
$$X_2 = X_\beta - A_2 \cdot |$$
$$C_2 \cdot X_\beta - X | \tag{12.11}$$
$$X_3 = X_\delta - A_3 . \left| C_3 \cdot X_\delta - X \right|$$
$$X(t+1) = \frac{X1(t) + X2(t) + X3(t)}{3}$$

The position vectors of α, β, and δ, and wolves, respectively, are X_α, X_β, and X_δ according to the above equation; the computations of A1, A2, and A3 are identical to A; and the computations of C1, C2, and C3 are identical to C. $D_\alpha = C_1 \cdot X_\alpha - X$, $D_\beta = C_2 \cdot X_\beta - X$, $D_\delta = C_3 \cdot X_\delta - X$ shows the distance between the three best wolves and the candidate wolves. Figure 12.3 shows that the candidate solution eventually falls inside the randomized circle described by α, β, and δ. The remaining contenders then modify their locations near the prey at random, guided by the best current

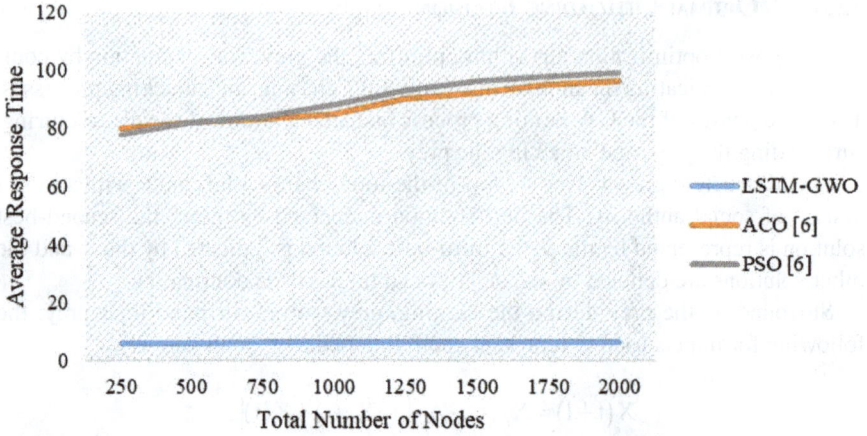

FIGURE 12.3 Average response time comparison with total number of nodes.

three wolves. They begin by searching for prey location information in a disorganized manner before focusing on assaulting the target.

12.5 RESULTS AND DISCUSSIONS

The proposed technique is built in MATLAB R-2020a to evaluate its effectiveness. The simulations were run on a PC with an Intel i5, 3.7GHz processor, and 8 GB of RAM. The tests are carried out to assess the suggested LSTM-GWO task offloading method in terms of many evaluation measures, including average delay/response time (RT), load degree of imbalance, and load imbalance standard deviation (SD). The degree of inequality, which is derived using the equation below, depicts the imbalance between the accessible fog nodes:

$$DI = \frac{Max(Rj)\ Min(Rj)}{Raverage}; j = 1,2,3\ldots\ldots\ldots N_{nodes}. \tag{12.12}$$

The standard deviation of offloaded task response times is being used to assess the load distribution across fog nodes, with a lower value indicating more balanced nodes, and is computed utilizing the formula below:

$$SD = \sqrt{\frac{\Sigma(R_{j-R_{average}})2}{Nmodes}}. \tag{12.13}$$

Figure 12.3 compares the suggested LSTM-GWO with PSO and ACO offloading techniques with the variable total number of nodes and displays average response times of offloaded tasks. In this case, when the number of tasks is expanded by adding additional sensors, the overall response time increases. In comparison to the ACO and PSO algorithms, the suggested LSTM-GWO offloading method maintains reduced average response times. LSTM-GWO average response time

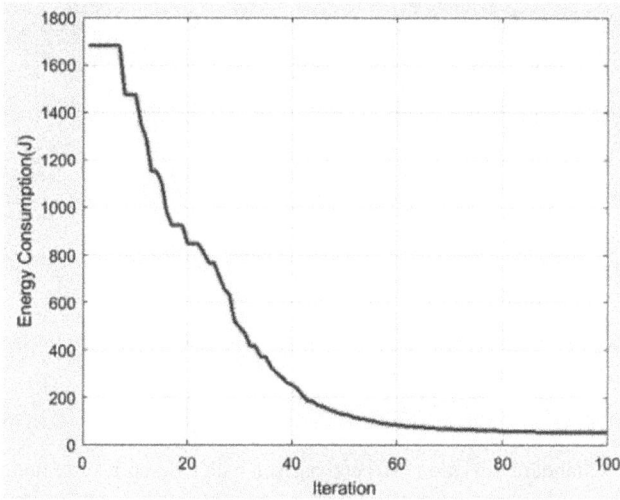

FIGURE 12.4 Energy consumption vs. iterations.

ranged from 3–6 seconds, as seen in the graph, but ACO and PSO's reaction times increased from 79–100 seconds as the number of nodes increased from 0–2000.

Figure 12.4 depicts the energy consumption as a function of iterations. The graph clearly shows that the method converges in a short number of iterations. Every iteration's energy consumption, measured in joules, is noted. The suggested LSTM-GWO method is faster than the PSO and ACO algorithms in terms of convergence: the optimal value is attained after iteration number 60. In comparison to the ACO and PSO algorithms, this finding indicated that the suggested LSTM-GWO algorithm investigated the search space and found an optimal value with an acceptable level of energy consumption.

Figure 12.5 displays the SD with respect to the number of fog nodes. The standard deviation of reaction times is the difference between the average response time and the response times of all offloaded tasks. When compared to ACO algorithms, suggested LSTM-GWO task offloading algorithm clearly increased SD.

Figure 12.6 depicts DI for the proposed LSTM-GWO with ACO and PSO algorithms, as well as the proposed method, as the number of fog nodes increases while maintaining the same parameter settings. The suggested LSTM-GWO algorithm retains fewer values, as shown in Figure 12.4. This indicates that the proposed technique efficiently balances the workload across the fog nodes.

12.6 CONCLUSION

The present work proposed and validated the task prediction and optimal offloading decision for FC cloud networks. Considering four evaluation parameters, average response time, energy consumption, load imbalance degree, and load imbalance SD as self-sufficient, adaptable, and knowledge-based sophisticated technologies and smart systems are now being created. Emergency/crisis management, aviation,

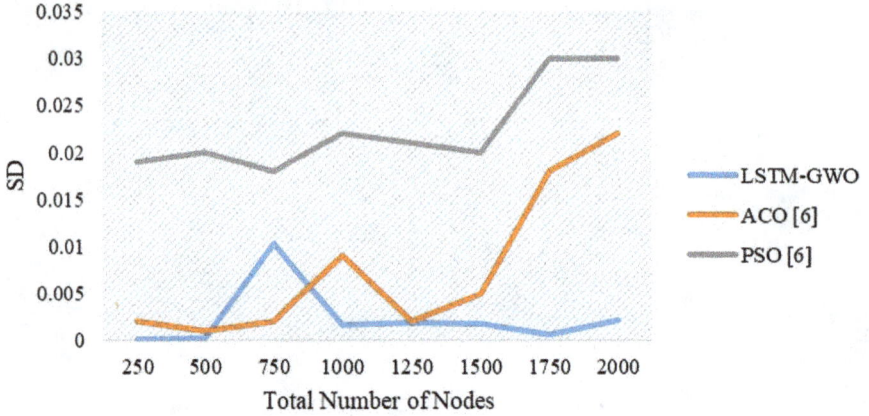

FIGURE 12.5 Standard deviation (SD) comparison with total number of nodes.

biotechnology, IoT, and mobile applications, to name a few, are revolutionizing the computer industry. The existing architecture of the centralized cloud has become unworkable due to workloads with a huge number of expanding devices. FC is quickly gaining traction as a distributed software model for having cloud-like applications close to end-users. It improves the computing capacities of network nodes and IoT products by delivering cloud-like computation and storage capabilities at reduced response times and energy consumption. FC also has the capacity to provide nodes, context-aware, dependability, and scaling. FC has been used to offload duties from apps offloading on end devices due to its many advantages. This enables quicker application execution by using the fog node's capabilities.

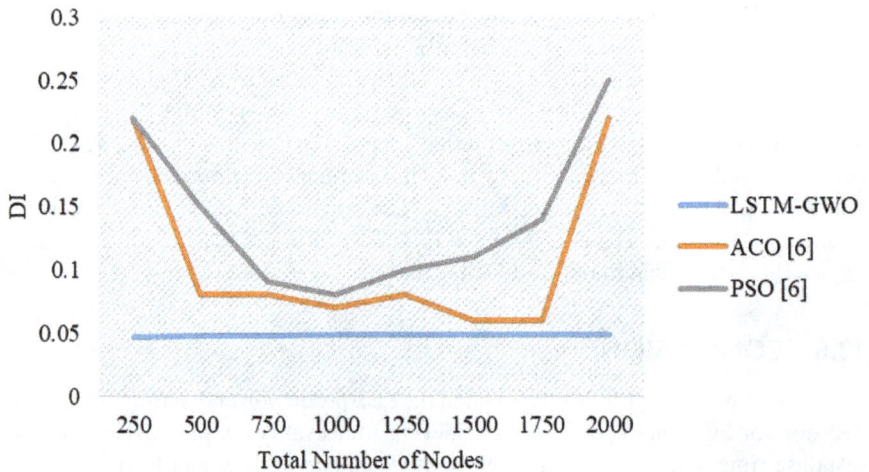

FIGURE 12.6 Degree of inequality (DI) comparison with total number of nodes.

Due to the dynamic environment of FC and many QoS characteristics, depending on the uses applied, the task offloading issue in FC is hard. The suggested method was simulated, and the results consist of parameters such as average response time, energy consumption, DI, and SD. The experimental results show that the proposed task prediction and optimal offloading decision algorithm provides a significant improvement in response times and effectively balances the tasks over the fog nodes. A formal model of task offloading on the fog nodes is provided. The LSTM-GWO task offloading algorithm clearly increased the standard deviation when compared to the specified algorithms. The DI for the proposed LSTM-GWO as compared to ACO and PSO algorithms retains fewer values as well.

REFERENCES

1. M. Aazam, S. U. Islam, S. T. Lone and A. Abbas, "Cloud of Things (CoT): Cloud-fog-IoT Task Offloading for Sustainable Internet of Things," IEEE Transactions on Sustainable Computing, vol. 7, pp. 87–98, 2020.
2. M. M. Razaq, B. Tak, L. Peng and M. Guizani, "Privacy-Aware Collaborative Task Offloading in Fog Computing," IEEE Transactions on Computational Social Systems, vol. 9, no. 1, pp. 88–96, Feb. 2022, doi: 10.1109/TCSS.2020.3047382.
3. Y. Jiang, Y. Chen, S. Yang and C. Wu, "Energy-Efficient Task Offloading for Time-Sensitive Applications in Fog Computing," IEEE Systems Journal, vol. 13, no. 3, pp. 2930–2941, Sept. 2019, doi: 10.1109/JSYST.2018.2877850.
4. Z. Zhou, H. Liao, B. Gu, S. Mumtaz and J. Rodriguez, "Resource Sharing and Task Offloading in IoT Fog Computing: A Contract-Learning Approach," IEEE Transactions on Emerging Topics in Computational Intelligence, vol. 4, no. 3, pp. 227–240, June 2020, doi: 10.1109/TETCI.2019.2902869.
5. D. Wang, Z. Liu, X. Wang and Y. Lan, "Mobility-Aware Task Offloading and Migration Schemes in Fog Computing Networks," IEEE Access, vol. 7, pp. 43356–43368, 2019, doi: 10.1109/ACCESS.2019.2908263.
6. M. K. Hussein and M. H. Mousa, "Efficient Task Offloading for IoT-Based Applications in Fog Computing Using ant Colony Optimization," IEEE Access, vol. 8, pp. 37191–37201, 2020.
7. Q. Wu, H. Liu, R. Wang, P. Fan, Q. Fan and Z. Li, "Delay-Sensitive Task Offloading in the 802.11p-Based Vehicular Fog Computing Systems," IEEE Internet of Things Journal, vol. 7, no. 1, pp. 773–785, Jan. 2020, doi: 10.1109/JIOT.2019.2953047.
8. H. Liao, Y. Mu, Z. Zhou, M. Sun, Z. Wang and C. Pan, "Blockchain and Learning-Based Secure and Intelligent Task Offloading for Vehicular Fog Computing," IEEE Transactions on Intelligent Transportation Systems, vol. 22, no. 7, pp. 4051–4063, July 2021, doi: 10.1109/TITS.2020.3007770.
9. Z. Zhou, H. Liao, X. Zhao, B. Ai and M. Guizani, "Reliable Task Offloading for Vehicular Fog Computing Under Information Asymmetry and Information Uncertainty," IEEE Transactions on Vehicular Technology, vol. 68, no. 9, pp. 8322–8335, Sept. 2019, doi: 10.1109/TVT.2019.2926732.
10. G. Zhang, F. Shen, Y. Yang, H. Qian and W. Yao, "Fair Task Offloading among Fog Nodes in Fog Computing Networks," 2018 IEEE International Conference on Communications (ICC), 2018, pp. 1–6, doi: 10.1109/ICC.2018.8422316.
11. M. Adhikari, M. Mukherjee and S. N. Srirama, "DPTO: A Deadline and Priority-Aware Task Offloading in Fog Computing Framework Leveraging Multilevel Feedback Queueing," in IEEE Internet of Things Journal, vol. 7, no. 7, pp. 5773–5782, July 2020, doi: 10.1109/JIOT.2019.2946426.

12. M. Aazam, S. Zeadally and K. A. Harras, "Offloading in Fog Computing for IoT: Review, Enabling Technologies, and Research Opportunities," Future Generation Computer Systems, vol. 87, pp. 278–289, 2018.

13. P. Kaur and S. Mehta, "Improvement of Task Offloading for Latency Sensitive Tasks in Fog Environment," In: Tiwari R., Mittal M., Goyal L.M. (eds) Energy Conservation Solutions for Fog-Edge Computing Paradigms. Lecture Notes on Data Engineering and Communications Technologies, vol 74. Springer, Singapore, 2022, doi: 10.1007/978-981-16-3448-2_3

14. X. Chen and G. Liu, "Energy-Efficient Task Offloading and Resource Allocation via Deep Reinforcement Learning for Augmented Reality in Mobile Edge Networks," IEEE Internet of Things Journal, vol. 8, no. 13, pp. 10843–10856, July1, 2021, doi: 10.1109/JIOT.2021.3050804.

15. P. Singh and R. Singh, "Energy-Efficient Delay-Aware Task Offloading in Fog-Cloud Computing System for IoT Sensor Applications," Journal of Network and Systems Management, vol. 30, p. 14, 2022, doi: 10.1007/s10922-021-09622-8

16. L. Yang, H. Zhang, X. Li, H. Ji and V. Leung, "A Distributed Computation Offloading Strategy in Small-Cell Networks Integrated with Mobile Edge Computing," IEEE/ACM Transactions on Networking (TON), vol. 26, no. 6, pp. 2762–2773, 2018

17. M. M. Bukhari and 1 T. M. Ghaza, "An Intelligent Proposed Model for Task Offloading in Fog-Cloud Collaboration Using Logistics Regression," Hindawi Computational Intelligence and Neuroscience Volume Article ID 3606068, p. 25, 2022, doi: 10.1155/2022/3606068

18. Z. Cao, P. Zhou, R. Li, S. Huang and D. Wu, "Multiagent Deep Reinforcement Learning for Joint Multichannel Access and Task Offloading of Mobile-Edge Computing in Industry 4.0," IEEE Internet of Things Journal, vol. 7, no. 7, pp. 6201–6213, July 2020, doi: 10.1109/JIOT.2020.2968951.

19. D. C. Nguyen, P. N. Pathirana, M. Ding and A. Seneviratne, "Privacy-Preserved Task Offloading in Mobile Blockchain With Deep Reinforcement Learning," IEEE Transactions on Network and Service Management, vol. 17, no. 4, pp. 2536–2549, Dec. 2020, doi: 10.1109/TNSM.2020.3010967.

20. Y. Kyung, "Performance Analysis of Task Offloading With Opportunistic Fog Nodes," IEEE Access, vol. 10, pp. 4506–4512, 2022, doi: 10.1109/ACCESS.2022.3141199.

21. S. Ming, T. Bao, D. Xie, H. Lv and G. Si, "Towards Application-Driven Task Offloading in Edge Computing Based on Deep Reinforcement Learning," Micromachines, vol. 12, no. 9, p. 1011, 2021, doi: 10.3390/mi12091011.

13 Balancing Innovation and Regulation
Navigating the Legal Frameworks of 5G Technology for Responsible and Sustainable Deployment

Rebant Juyal

13.1 INTRODUCTION

The advent of 5G technology has been hailed as a game-changer and a significant milestone in telecommunications. With faster speeds, lower latency, and greater connectivity, 5G promises to revolutionize the way we live, work, and interact with each other. It is expected to support a range of emerging technologies, such as autonomous vehicles, smart cities, and the Internet of Things (IoT), making them more efficient and effective. However, deploying 5G technology also raises important legal issues and concerns that need to be addressed. The importance of laws governing 5G technology cannot be overstated, as they play a crucial role in ensuring technology's safe and responsible deployment [1–3].

The legal framework provides the necessary guidelines, regulations, and standards to govern the use of 5G technology, safeguarding the interests of users, businesses, and the public [4]. The significance of laws governing 5G technology lies in their ability to address various legal issues like data privacy and security concerns, liability in case of accidents or failures, intellectual property rights and licensing issues, environmental concerns, and competition law issues. Without proper legal guidelines, deploying 5G technology could lead to various legal and ethical issues that could impact its adoption and growth. The legal framework governing 5G technology is also essential for ensuring a level playing field for all stakeholders [5]. Regulations and guidelines are necessary to prevent anti-competitive behavior, protect intellectual property rights, and ensure fair and equitable access to 5G networks and services. It is also critical to ensure that the deployment of 5G technology is consistent with social and environmental values [6].

The deployment of 5G technology has led to a new set of legal and regulatory issues that need to be addressed. The legal framework of 5G technology is complex, involving national and international laws and regulations, and the role of governments

DOI: 10.1201/9781003407836-13

in regulating the deployment and operation of 5G networks is crucial. This article will provide an overview of the laws governing 5G technology, the role of governments in regulating it, and the legal issues that may arise from its deployment [7].

13.2   OVERVIEW OF 5G TECHNOLOGY

5G technology is the latest and most advanced mobile network technology by contemporary standards. It represents a significant upgrade from its predecessor, 4G technology, and offers faster download and upload speeds, lower latency, and higher bandwidth. 5G is designed to support diverse applications and use cases, from smartphones to IoT, and mission-critical communication systems. One of the most significant advantages of 5G technology is its speed. 5G networks can deliver peak download speeds of up to 20 gigabits per second (Gbps), approximately 100 times faster than 4G. This means that users can download large files, stream high-quality video, and perform other data-intensive tasks more quickly and efficiently than ever before [8–9].

Another key feature of 5G technology is its low latency. Latency refers to the time it takes for data to travel between a device and the network. 5G networks can deliver latency as low as 1 millisecond (ms), which is significantly faster than the 50 ms latency offered by 4G. This reduced latency is significant for applications that require real-time communication, such as remote surgery, self-driving cars, and virtual reality gaming. 5G technology also offers increased bandwidth, which means that more devices can connect to the network simultaneously without experiencing a decrease in speed or performance. This is particularly important for the IoT, which requires a large number of devices to be connected to the network at the same time [10].

5G technology is also designed to be more efficient than 4G, which means it uses less energy to transmit data. This feature is essential for several reasons, including reducing mobile networks' environmental impact and increasing mobile device battery life. Overall, 5G technology represents a significant step forward in mobile network technology, offering faster speeds, lower latency, and higher bandwidth than its predecessor. It can potentially revolutionize a wide range of industries and applications, from healthcare to transportation to entertainment, and is poised to become the backbone of the digital economy [11].

13.3   EXPLANATION OF 5G TECHNOLOGY AND ITS FEATURES

5G technology is the fifth generation of wireless technology designed to enhance mobile networks' speed, reliability, and connectivity. Unlike its predecessors, 5G technology uses higher frequency bands, including the millimeter-wave (mm-wave) spectrum, allowing faster data transmission rates, lower latency, and greater capacity. One of the main features of 5G technology is its speed. With speeds up to 100 times faster than 4G, 5G technology allows for ultra-fast downloads and uploads, making it ideal for streaming high-definition content, virtual and augmented reality applications, and online gaming [12].

Low latency also stands as another central feature of 5G technology. Latency refers to the time it takes for data to travel from the sender to the receiver. With 5G technology, latency is reduced to less than 1 ms, so there is almost no delay in transmitting data, enabling real-time communication and interaction. 5G technology also has greater capacity than its predecessors. With more bandwidth, 5G networks can support a more significant number of devices, making it possible for many devices to connect to the network simultaneously without compromising the speed or quality of onection. 5G technology additionally enables network slicing, allowing service providers to divide their networks into smaller, virtual ones. Each network slice can be customized to meet the specific needs of different applications, allowing for greater flexibility and efficiency [13].

13.3.1 COMPARISON OF 5G WITH PREVIOUS TECHNOLOGIES

5G technology offers substantial advantages over previous technologies, including faster speeds, lower latency, greater capacity, and enhanced reliability [14]. Compared to its predecessors, the following advantages can be postulated of 5G technology:

- Faster speeds: 5G technology can deliver speeds up to 100 times faster than 4G, enabling faster downloads and uploads.
- Lower latency: With 5G technology, latency is reduced to less than 1 ms, enabling real-time communication and interaction.
- Greater capacity: 5G networks have greater bandwidth and can support a larger number of devices, making it possible for many devices to connect to the network simultaneously without compromising the speed or quality of the connection.
- Enhanced reliability: 5G technology uses advanced technologies such as beamforming and network slicing to enhance the reliability and performance of the network.
- Support for emerging technologies: 5G technology is designed to support emerging technologies such as autonomous vehicles, IoT, and smart cities, enabling them to operate more efficiently and effectively.

13.3.2 APPLICATION AND BENEFITS OF 5G TECHNOLOGY

5G technology benefits a wide range of industries and applications, enabling the development of new and innovative technologies having the potential to improve human lives and enhance ability to communicate and interact within the society. Its potential features make it ideal for emerging technologies like

- IoT: 5G technology is designed to support a massive number of connected devices, making it ideal for IoT applications such as smart homes, smart cities, and industrial automation.
- Autonomous vehicles: The low latency and high reliability of 5G technology make it an ideal platform for autonomous vehicles, enabling real-time communication between vehicles and infrastructure.
- Healthcare: 5G technology can improve access to healthcare services, including telemedicine and remote patient monitoring, by enabling real-time communication between patients and healthcare providers.

- Entertainment: The high speed and capacity of 5G technology make it ideal for streaming high-definition content, virtual and augmented reality applications, and online gaming.
- Manufacturing: 5G technology can enhance the efficiency and productivity of manufacturing processes by enabling real-time communication between machines and devices.

13.4   LAWS GOVERNING 5G TECHNOLOGY

Several national and international laws and regulations govern the deployment and operation of 5G technology. Some of the key laws and regulations include

- International Telecommunication Union (ITU) regulations: The ITU is a specialized agency of the United Nations that sets international standards for telecommunications. The ITU has developed standards for 5G technology, including technical specifications for 5G networks and devices.
- National telecommunications laws: Each country has its own laws and regulations governing telecommunications. These laws address spectrum allocation, licensing, and network infrastructure issues.
- Data protection laws: The deployment of 5G technology involves collecting, processing, and storing large amounts of data. Data protection laws regulate how this data is collected, processed, and stored.
- Environmental laws: The deployment of 5G networks requires the installation of new infrastructure, such as antennas and base stations. Environmental laws regulate the installation of this infrastructure, including issues such as land use and radiation emissions.

13.4.1   ROLE OF GOVERNMENT IN REGULATING 5G TECHNOLOGY

The role of the government in regulating the deployment and operation of 5G technology is crucial. Governments are responsible for ensuring that the deployment of 5G networks is safe, secure, and reliable [15]. Some of the vital roles of governments in regulating 5G technology include

- Spectrum allocation: Governments allocate spectrum for the operation of 5G networks. This allocation is done through auctions or licensing processes.
- Licensing: Governments issue licenses to network operators to operate 5G networks. These licenses may include conditions such as coverage obligations and quality of service requirements.
- Security: Governments are responsible for ensuring the security of 5G networks. This includes ensuring that networks are protected from cyberattacks and other security threats.
- Environmental protection: Governments are responsible for ensuring that the deployment of 5G networks complies with environmental laws and regulations.

13.4.2 Legal Issues Arising from the Deployment of 5G Technology

As with any new technology, 5G raises a number of legal issues that need to be addressed. Some of the key legal issues in 5G technology include data privacy and security concerns, liability issues in case of accidents or failures, intellectual property rights and licensing issues, environmental concerns and impact on health, and competition law issues.

a. Data privacy and security concerns: 5G networks will generate large amounts of data, which will be transmitted over the network and stored in data centers. This data can include personal information, such as location data, browsing history, and communication records. As such, there are concerns around data privacy and security, and how this data will be protected from unauthorized access or use. Consequently, ensuring due privacy of an individual and security of the data remains a critical concern in the realms of law and administration.

b. Liability issues in case of accidents or failures: As with any complex technology, there is a risk of accidents or failures in 5G networks. Further deployment of 5G networks involves installation of new infrastructures like antennas and base stations. Liability concerns may arise in case of any failures. If infrastructure deployed for the technology on account of some failure causes harm to individuals or property, that may patently lead to liability issues for telecom companies or equipment manufacturers in such cases.

c. Intellectual property rights and licensing issues: The development of 5G technology involves a range of intellectual property rights, including patents, trademarks, and copyrights. There are also licensing issues related to the use of 5G technology, including the licensing of standard essential patents.

d. Environmental concerns and impact on health: There are concerns around the environmental impact of 5G technology, including the energy consumption of data centers and the use of rare earth metals in the production of equipment. There are also concerns around the potential impact of 5G on human health, although the scientific consensus is that there is no evidence of harm from exposure to radiofrequency radiation from 5G networks.

e. Competition law issues in 5G technology: The development of 5G networks will have significant implications for competition in the telecom industry. There are concerns around the potential for dominant players to emerge and stifle competition, as well as concerns around the potential for anti-competitive behavior in the licensing of intellectual property rights.

5G technology is also governed by a set of laws and regulations. Some of the key laws and regulations that apply in India include

a. Indian Telegraph Act, 1885: The Indian Telegraph Act, 1885 is a key piece of legislation governing telecommunications in India. This act provides for the establishment, maintenance, and operation of telecommunications networks and services in India. The act also provides for the allocation of spectrum and licensing of telecommunications services.

b. Telecom Regulatory Authority of India (TRAI) regulations: The TRAI is the regulator for the telecommunications industry in India. The TRAI has issued regulations on a range of issues related to telecommunications, including spectrum allocation, licensing, and quality of service.

c. Information Technology (IT) Act, 2000: The IT Act, 2000 is a key piece of legislation governing the use of technology in India. The act provides for the regulation of electronic communication, including data protection and cybersecurity.

d. National Digital Communications Policy, 2018: The National Digital Communications Policy, 2018 is a policy framework that sets out the government's vision for the telecommunications sector in India. The policy outlines a range of measures to promote the deployment and adoption of digital technologies, including 5G.

The Indian government has also taken steps to regulate the deployment and operation of 5G networks in the country. For example, the government has set up a committee to study and recommend issues related to 5G deployment, including spectrum allocation and network infrastructure. The government has also issued guidelines on installing telecom infrastructure, including 5G infrastructure, to ensure environmental laws and regulations compliance. Legal issues arising from deploying 5G technology in India include data privacy, liability, health and safety, and competition. For example, concerns have been raised about the potential impact of 5G technology on public health, particularly concerning exposure to radiation from the network infrastructure. Regulators and network operators must ensure that the deployment of 5G networks in India is safe for the public.

13.5 JUDICIAL JUDGMENTS

There have been several judicial judgments in India related to the allocation and pricing of spectrum for telecommunications services, including 5G. Some of the key judgments are

a. 2G Spectrum Case: In 2012, the Supreme Court of India cancelled 122 2G spectrum licenses that had been granted by the government in 2008. The court held that the government had not followed proper procedures for allocation and pricing the spectrum, and that the allocation had resulted in a loss of revenue for the government. The judgment had significant implications for the telecom industry in India, including the cancellation of licenses of several telecom companies.

b. Airtel and Vodafone-Idea case: In 2020, the Supreme Court of India upheld the government's definition of adjusted gross revenue (AGR) for calculating telecom companies' license fees and spectrum usage charges. The court's decision had significant financial implications for telecom companies, including Bharti Airtel and Vodafone-Idea, who were required to pay billions of dollars in outstanding dues to the government.

c. Telecom Regulatory Authority of India v. Association of Unified Telecom Service Providers of India: In 2015, the Supreme Court of India upheld the TRAI's powers to fix the reserve price for spectrum auctions. The court held that the reserve price could be set based on market conditions and that the TRAI was the appropriate body to determine the reserve price. The judgment had significant implications for spectrum pricing and allocation in India, as it affirmed the TRAI's role in regulating the telecom industry.

d. Vodafone Idea Ltd. v. Union of India: In 2021, the Supreme Court of India directed the government to consider granting a 10-year moratorium on payment of AGR-related dues by telecom companies. The court's decision was seen as a relief for telecom companies, including Vodafone-Idea, which had been struggling to pay their outstanding dues to the government.

e. These judgments highlight the importance of proper procedures for allocation and pricing of spectrum for telecom services, including 5G. The judgments also emphasize the role of regulatory bodies, such as the TRAI, in overseeing the telecom industry and setting policies and regulations for spectrum allocation and pricing.

13.5.1 GLOBAL PERSPECTIVES ON 5G TECHNOLOGY AND LAW

The deployment of 5G technology is a global phenomenon; as such, laws and regulations related to 5G vary significantly between different countries. While some countries have taken a proactive approach to regulating 5G, others are still developing a legal framework for the technology. For example, in the United States, the Federal Communications Commission (FCC) has been actively involved in the development of 5G, including the allocation of spectrum and the regulation of equipment manufacturers. The government has played a central role in developing 5G in China, with significant investments in research and development and deploying 5G networks.

Europe focuses on developing a common approach to regulating 5G, with the European Union (EU) taking a lead role in coordinating regulations across member states. The EU has also taken steps to address some critical legal issues related to 5G, including data privacy and security concerns, through the General Data Protection Regulation (GDPR). Despite these efforts, there are significant challenges in harmonizing a global legal framework for 5G technology. One key challenge is the differing priorities and approaches of different countries. For example, some countries priorities national security concerns, while others priorities data privacy or competition law concerns.

There are also challenges related to the pace of technological development, which can make it difficult for regulators to keep up with changes in the industry. This is particularly true in the case of 5G, which is evolving rapidly and is expected to impact a range of industries significantly.

International organizations, such as the International Telecommunication Union (ITU) and the World Health Organization (WHO), play a crucial role in regulating 5G technology. The ITU is responsible for coordinating global efforts to allocate spectrum for 5G, which is a critical aspect of ensuring the successful deployment of the technology. Spectrum allocation is necessary to ensure that different network providers can access the frequency bands to operate 5G networks effectively. The ITU also develops technical standards related to 5G technology. These standards are essential in ensuring that different components of the 5G ecosystem, such as network infrastructure and mobile devices, are interoperable and can work seamlessly together. This is critical to the success of 5G, as it enables users to access the benefits of the technology regardless of their location or service provider.

In addition to spectrum allocation and technical standards, international organizations also play an essential role in addressing concerns related to the health and safety implications of 5G technology. The WHO, for example, is responsible for assessing the potential health risks associated with exposure to electromagnetic fields, including those generated by 5G networks. The WHO's International EMF Project, a collaboration between WHO and ITU, developed guidelines for the safe usage of wireless technologies. These guidelines are intended to help governments, industry, and the public assess and manage potential health risks associated with exposure to electromagnetic fields. International organizations such as the ITU and the WHO provide collaboration and knowledge-sharing platforms between countries and stakeholders. This is particularly important in the case of 5G, a global technology expected to impact a range of industries and sectors significantly.

The above analysis certifies that regulating 5G technology is a complex and multifaceted issue with significant implications for various stakeholders. While there are challenges in harmonizing global regulations for the technology, it is clear that a coordinated and collaborative approach is necessary to ensure that 5G is deployed responsibly and sustainably.

13.6 CONCLUSION

The introduction of 5G networks is a significant milestone in the telecommunications industry, providing unprecedented speed, capacity, and connectivity to users. 5G networks are expected to power a range of innovative technologies, including autonomous vehicles, smart cities, and IoT. However, the deployment and use of 5G networks also pose significant legal and regulatory challenges, including issues related to spectrum allocation, privacy, security, and environmental impact. To ensure the responsible and equitable deployment of 5G networks, legal frameworks are necessary to regulate their development and use.

The exploratory research of the article certifies prevalence of critical legal issues concerning the deployment of the 5G technology in the telecom sector of India. While the Indian republic is a free market economy, yet the obligation of the states in protection and preserving the rights and interests of Indian citizens including business and others stands imperative. In this regard, instituting appropriate legal instruments to deal with challenges such as environmental and health concerns, data privacy issues, liability concerns, among others. It is critical to address the legal implications and challenges that come with the continued development and deployment of 5G technology. It is essential to adapt legal frameworks to keep pace with technological advancements and ensure that the benefits of 5G technology are realized in a responsible and sustainable manner. With the future outlook of 5G technology looking promising, there will be ongoing legal implications that need to be addressed to mitigate any potential risks associated with it.

This chapter attempts to highlight the importance of finding a balance between innovation and regulation in 5G technology. The analysis of the article provides a strong inclination towards the idea that by addressing the legal implications and challenges associated with 5G technology, the state can ensure that benefits of this technology are realized while minimizing its potential risks.

REFERENCES

1. Qualcomm. (2023). What is 5g: Everything you need to know about 5G. Qualcomm. https://www.qualcomm.com/5g/what-is-5g.
2. Sehgal, D. R. (2020, September 11). 5G: Economic opportunity and concerns. *iPleaders*. https://blog.ipleaders.in/5g-economic-opportunity-concerns/.
3. Popescu, M. P.-A., & Crețu, C. (2020, June 25). Legal challenges in implementing the 5g EU toolbox and potential damaging effects on electronic communication providers and consumers. Lexology. https://www.lexology.com/library/detail.aspx?g=f87c09e8-d1c2-4930-8217-9141ee3b330f.
4. Singh, M. Advent of 5G: Opportunities and challenges. AKGEC International Journal of Technology, 7(2), 1–8.
5. Bagnall, M., & Hall, K. (2022, August 19). Non-terrestrial mobile networks – the final frontier for 5G? Lexology. https://www.lexology.com/commentary/tech-data-telecoms-media/international/wiggin-llp/non-terrestrial-mobile-networks-the-final-frontier-for-5g
6. Parikh, Dr. M., Parikh, V., Saha, P., & Sarangal, S. (2020). 5G Technology in India: Strategic, legal and regulatory considerations (pp. 1–30). Nishith Desai Associates. https://www.nishithdesai.com/fileadmin/user_upload/pdfs/Research_Papers/5G-Technology-in-India.pdf.
7. Simpson, T. W. and Simpson, D. (2019, September 3). Why 5G requires new approaches to cybersecurity. *Brookings*. https://www.brookings.edu/research/why-5g-requires-new-approaches-to-cybersecurity/.
8. 5G - Fifth generation of mobile technologies. (2022, April). ITU. https://www.itu.int/en/mediacentre/backgrounders/Pages/5G-fifth-generation-of-mobile-technologies.aspx.
9. World Health Organization. "WHO Fact Sheet on Electromagnetic Fields and Public Health." WHO, Oct. 2014, https://www.who.int/news-room/fact-sheets/detail/electromagnetic-fields-and-public-health-mobile-phones.
10. Humayun, M., Hamid, B., Jhanjhi, N. Z., Suseendran, G., & Talib, M. N. (2021). 5g network security issues, challenges, opportunities and future directions: A survey. Journal of Physics: Conference Series, 1979(1), 012037. https://doi.org/10.1088/1742-6596/1979/1/012037.
11. Robles-Carrillo, M. (2021). European union policy on 5G: Context, scope and limits. Telecommunications Policy, 45(8), 102216. https://doi.org/10.1016/j.telpol.2021.102216.

12. Nissenbaum, H. (2010). Privacy in context: Technology, policy, and the integrity of social life. Stanford University Press.
13. Brouwer, J., & Koops, B. J. (2018). The ethics of ambient intelligence. Science and Engineering Ethics, 24(3), 693–706.
14. European Commission. (2018). Liability for emerging digital technologies. European Commission.
15. Werbach, K. (2019). The Blockchain and the new architecture of trust. MIT Press.

Index

S

SCMA, 67
SDMA, 91
Security, 32, 224
Sensing, 73
Self-driving vehicles, 65
Self-optimization, 37
SINR, 55
Small-base stations, 38
SNR, 45
Software-defined networks, 198
Software-defined radio, 79
Spectral efficiency, 33, 51, 80
Stochastic geometry, 81
Sub-6 GHz, 172, 179
Successive interference cancellation, 16, 200
Sum-rate, 96
Superposition coding, 68
Sustainable deployment, 221
SWIPT, 51, 52
System reliability, 23

T

Task prediction, 207, 208
TDMA, 65
Termination, 28
Throughput, 73
Time switching relaying, 52

Traffic, 37, 38, 65
Transmissions, 71, 77, 79
Transmit antenna selection, 78

U

UFMC, 163, 164
Unmanned aerial vehicles, 80
Urban Transformation, 179, 188
URLLC, 111
User clustering, 24, 27, 28
User experience, 37, 91, 135
User grouping, 22, 23
User pair power allocation, 76

V

Vehicle-to-vehicle, 65
Visible light communications, 198
Vital, 2, 165, 196
VLC-NOMA, 124

W

WiMAX, 13, 153
Wireless, 2, 15, 18, 19
Wireless communication systems, 2, 14, 16
Wireless personal area network, 77
Wireless sensor networks, 82
Wireless technologies, 1, 39, 40

For Product Safety Concerns and Information please contact our EU
representative GPSR@taylorandfrancis.com
Taylor & Francis Verlag GmbH, Kaufingerstraße 24, 80331 München, Germany

www.ingramcontent.com/pod-product-compliance
Lightning Source LLC
Chambersburg PA
CBHW060402220326
41598CB00023B/2992